HANDBOOK
OF
PACKAGE MATERIALS

some other AVI books

Food Science and Technology

BEVERAGES: CARBONATED AND NONCARBONATED   *Woodroof and Phillips*

BREAD SCIENCE AND TECHNOLOGY   *Pomeranz and Shellenberger*

CEREAL SCIENCE   *Matz*

CEREAL TECHNOLOGY   *Matz*

COMMERCIAL FRUIT PROCESSING   Cloth and Soft Cover   *Woodroof and Luh*

COMMERCIAL VEGETABLE PROCESSING   Cloth and Soft Cover   *Luh and Woodroof*

COOKIE AND CRACKER TECHNOLOGY   *Matz*

ENCYCLOPEDIA OF FOOD ENGINEERING   *Hall, Farrall and Rippen*

ENCYCLOPEDIA OF FOOD TECHNOLOGY   *Johnson and Peterson*

FABRICATED FOODS   *Inglett*

FOOD AND THE CONSUMER   Soft Cover   *Kramer*

FOOD COLORIMETRY: THEORY AND APPLICATIONS   *Francis and Clydesdale*

FOOD FOR THOUGHT   Soft Cover   *Labuza*

FOOD PACKAGING   *Sacharow and Griffin*

FOOD PROCESS ENGINEERING   Cloth and Soft Cover   *Heldman*

FOOD PRODUCTS FORMULARY, VOL. 1   *Komarik, Tressler and Long*   VOL. 2   *Tressler and Sultan*   VOL. 3   *Tressler and Woodroof*

LABORATORY MANUAL FOR FOOD CANNERS AND PROCESSORS, 3RD EDITION, VOLS. 1 AND 2   *National Canners Association*

MICROWAVE HEATING, 2ND EDITION   *Copson*

PACKAGE PRODUCTION MANAGEMENT, 2ND EDITION   Cloth and Soft Cover   *Raphael and Olsson*

POTATO PROCESSING, 3RD EDITION   *Talburt and Smith*

PRACTICAL MEAT CUTTING AND MERCHANDISING, VOLS. 1 AND 2   Soft Cover   *Fabbricante and Sultan*

PRINCIPLES OF PACKAGE DEVELOPMENT   *Griffin and Sacharow*

PROCESSED MEATS   *Kramlich, Pearson and Tauber*

RHEOLOGY AND TEXTURE IN FOOD QUALITY   *deMan, Voisey, Rasper and Stanley*

SNACK FOOD TECHNOLOGY   *Matz*

TECHNOLOGY OF WINE MAKING, 3RD EDITION   *Amerine, Berg and Cruess*

THE FREEZING PRESERVATION OF FOODS, 4TH EDITION, VOLS. 1, 2, 3 AND 4   *Tressler, Van Arsdel and Copley*

THE TECHNOLOGY OF FOOD PRESERVATION, 3RD EDITION   *Desrosier*

HANDBOOK
OF
PACKAGE MATERIALS

by STANLEY SACHAROW, B.A.,M.A.

Professional Member,
Packaging Institute

THE AVI PUBLISHING COMPANY, INC.
WESTPORT, CONNECTICUT

© *Copyright 1976 by*
THE AVI PUBLISHING COMPANY, INC.
Westport, Connecticut

Library of Congress Catalog Card Number: 76-27465
ISBN-0-87055-207-4

Printed in the United States of America

Dedicated to
The Packaging Industry—Past and Present

Foreword

For the many people with varied backgrounds who are active in packaging, the name Stanley Sacharow is readily associated with worthwhile books on this fascinating subject. These same people would agree that the presentation of pertinent information on packaging in general, and packaging materials in particular, is sorely needed. The quantity of definitive books dealing with the science and engineering of packaging does not, at this time, measure up to the challenge and complexity of the field. Packaging predates history while being as new as tomorrow's innovative development. It is always changing and dogged in its constancy. In order to meet the myriad of challenges presented by packaging, our educational system must provide properly trained personnel. A critical ingredient in the educating process is textbooks on packaging. This book provides an important contribution to the glaring shortage of comprehensive texts.

We are actively pursuing education and research in Packaging Science and Engineering at Rutgers University. In order to provide a much needed informative reference for the field of packaging science and package engineering, an insight into the very foundation of packaging, namely packaging materials, is required at Rutgers, at other educational institutions and throughout the industry. This text will be extremely useful in filling the existing gap.

Not only do I say thanks Stan for a job well done, but also, keep up the good work, because it can only benefit packaging.

DARRELL R. MORROW, PH.D.
Chairman,
Packaging Science and Engineering,
Rutgers University

January 1, 1976

Preface

This book summarizes and attempts to classify the various materials used for the protection and sale of most consumer packages. An important point to note is that primary packaging materials have not been covered—other than those that are common to both primary and secondary units. While primary materials such as wood crates, plastic barrels, etc., represent a major industry, their discussion could be the subject of a subsequent book dealing with industrial packaging. There has long been needed a basic text solely devoted to packaging materials. A researcher demands a volume that is capable of providing some basic data for his activities. It is hoped that this text will serve this purpose.

Packaging materials represent a major stabilizing force in an ever changing industry. Although new concepts and machines appear frequently, many of the most widely used materials have been around for decades. Prior to the onset of the "Plastics Age," the "big three" substrates in flexible packaging were cellophane, aluminum foil and paper. These still constitute, as they have since the early 1930's, the largest volume of materials used to prepare flexible packages. Polyethylene was an addition to the group in the early 1950's and with the previous materials, those materials now constitute the "big four" in the flexible packaging industry. New materials are always on the drawing board and perhaps in the not too distant future a revolutionary concept or substrate will appear.

Credit in writing a book must be given to many people and to many companies. To Dr. Aaron L. Brody of Mead Packaging, my gratitude for reviewing the text prior to publication. His suggestions and comments have served to bring into focus the full scope of this material. To Dr. Robert F. Testin, who has again, as in the past, prepared an outstanding chapter on the most important subject of package waste in cooperation with Mr. Gilbert F. Bourcier. Their contribution is deeply appreciated. Mr. Dixie A. Dean of Fisons, Ltd., Pharmaceutical Division has graciously given me permission to use his data on closures (Chap. 2). His cooperation is truly an example of international friendship. To Dr. Darrell A. Morrow of Rutgers University, my hope that the information in the book will serve to educate his many students. He has done an outstanding job at Rutgers in preparing students for industrial positions and I appreciate both his reviewing and writing the Foreword to this text. To my secretary, Jill S. Vitale, my deepest appreciation for the time she devoted to typing the manuscript. It could not have been done without her cooperation. And most important of all, to my wife, Beverly Lynn, and sons, Scott Hunter and Brian Evan, my thanks for understanding the value of this effort. Without their cooperation this book could not have been written.

STANLEY SACHAROW

September 1, 1976

Contents

CHAPTER   PAGE

1.   ORIGIN OF PACKAGING DEVELOPMENT   1
2.   GLASS PACKAGING AND CLOSURES   17
3.   METALS   41
4.   PLASTICS   65
5.   PAPER, PAPERBOARD AND CORRUGATED
      FIBERBOARD   84
6.   FLEXIBLE PACKAGING—PAPER   99
7.   FLEXIBLE PACKAGING—FILMS 108
8.   FLEXIBLE PACKAGING—ALUMINUM FOIL 169
9.   FLEXIBLE PACKAGING MATERIAL CONVERTING . . 182
10.   PACKAGING AND THE ENVIRONMENT[1] 206
      INDEX 241

[1] By Robert F. Testin, Ph.D.

CHAPTER 1

Origin of Packaging Development

INTRODUCTION

The use of packages and containers goes back to the dawn of history. In prehistoric times early man must have carried water in something, perhaps a crude cup shaped from a large leaf, a clam shell, or a joint from a bamboo stem. His first man-made package or container may easily have been a clay bowl. Man's use of fire was followed by his learning to bake clay vessels, but even before that he may have simply left the crudely shaped bowls and pots to dry in the sun.

Even the more modern materials which we know today have early beginnings. Glass bottles are known to have been in use in Egypt more than four thousand years ago. The pyramids themselves were containers deluxe. Moisture-proof, tamper-proof, impervious to light, they preserved their contents essentially in original form for thirty or more centuries.

Greek artisans produced pottery of exquisite beauty. Jars, vases, urns, pitchers and bottles have been recovered from buried cities; decorations on them are perfectly preserved, and in some even the grain or wine that they held is still identifiable. Bronze and the other early metals became important packaging materials in many civilizations. Many of these also have survived virtually undamaged. Grain chests, burial urns, wine bottles and cooking vessels give ample evidence of the craftsmanship of their makers. Package and container manufacture in ancient times was indeed an art.

Later Developments

The cotton drawstring pouch is a good example of a later development in packaging. Cloth was undoubtedly used thousands of years ago. However, cloth rots and disintegrates rapidly unless it is protected from moisture, so we have little if any evidence of its use in pouches or bags.

Larger bags and sacks were a familiar sight in the old general store. They were usually made of burlap, a cloth woven from jute or hemp fibers. This coarse material was also called crocus; from that name came the "croker" sack of the southern United States. As well as being strong and inexpensive, burlap had the advantage of an open weave which allowed free passage of air and moisture. This made the

1

Courtesy of Walter Landor and Associates

FIG. 1.1. EARLY CEREAL PACKAGE

Such packages were commonplace in the early 20th century.
Even today this type of container is widely used for oats.

burlap bag ideal for bulk storage of potatoes, onions, and other such vegetables that would otherwise rot.

Along with cloth paper is one of the oldest man-made materials. However, as a container material paper is somewhat more recent. Only in comparatively modern times were waterproof papers developed. Before this invention, although paper was a common packaging or wrapping medium, it was not very satisfactory as a container.

For dry products the paper bag is in universal use. The brown or kraft paper bag is so common today that it's hard to believe its history is so short. Not until the early years of the 20th century did the kraft bag appear; the square-bottomed "set-up" bag came along even later.

Preformed paperboard boxes are also a 20th century development. Until then they were available only as expensive merchandise such as jewelry. Essentially all of the packages that we accept as commonplace today are a result of the industrial revolution of the late 19th and early 20th centuries. Without the mass production methods developed during this era, our modern packages would be expensive and completely unavailable to the average purchaser.

Glass bottles are quite modern. Glass has been known and used for thousands of years, but until comparatively recent times glass containers were essentially a work of art. They were individually hand blown, frequently exquisitely decorated and quite fragile. While they were ideally suited to contain expensive perfumes, for example, they were hardly practical for dispensing cheap merchandise. As with paper, industrial production can be credited with making the glass bottle and jar available for even the least expensive goods. Today the "one-way" glass bottle is actually cheaper to produce new than it is to clean and re-use.

Quite a different situation exists with wood. Plentiful and cheap a few generations or so ago, wood as a container has been largely replaced except for highly specialized use. Cases for bananas, melons, and other produce are still made of cheap wood slats, and there are other limited applications. But wood today is expensive, hard to form, and does not lend itself to the attractive, modern containers so much in demand. Wood was once in great demand for barrels, casks, large cases, coffins, and other sturdy containers. Charred white oak barrels are still very important for aging whiskey, but this is a specialized application that depends on the ability of the charred wood to absorb undesirable elements in the green liquor. Cigar boxes traditionally were made of fine, beautifully grained woods such as Philippine mahoghany. Most of these have now been replaced with paperboard. It can be said that most of the traditional materials used in packaging for the last four thousand years are rapidly being replaced by materials that were hardly known fifty years ago.

Modern Improvements

The modern materials of packaging include paper, paperboard, steel, aluminum, glass, wood, textiles, and plastics. Of these, aluminum and plastics have been a factor in the packaging industry for only thirty years or so. The others, while far from new to the industry, have undergone such developments and changes that they are almost new materials as far as their characteristics are concerned.

Courtesy of Walter Landor and Associates
FIG. 1.2. VARIOUS TIN CANNISTERS—LATE 19TH CENTURY
Note the extremely interesting designs on these lithographed tin cans for cocoa packages.
These items are now collectible.

Packaging papers, for example, are now available in at least fifteen broad categories, from the lightest tissues to the heaviest krafts, coated, treated, and laminated, to produce hundreds of specialty items.

Steel, traditionally the leader in materials for rigid containers, has changed too. For years tin-coated steel in the form of "tin" cans packaged essentially all of the world's canned goods. During World War II tin became unavailable. There simply is no source of native tin in the United States. Now can makers have developed tin-free steel cans with various organic linings. They may be cheaper and they eliminate the soldered side seam, thus allowing all-around printing for greater eye appeal.

The synthetic materials, known more familiarly as plastics, are truly a "modern improvement." The first synthetic, celluloid, was discovered more than one hundred years ago, but only since World War II have the plastics really affected the packaging industry. Perhaps the most commonly known plastic is cellophane, which is not a synthetic material since it is made from natural cellulose. It frequently serves as a base for laminating with many of the true synthetics to enhance its own naturally desirable qualities.

Certain plastics available today have names that have already become household words. Perhaps nylon is the best example. First brought to the public's attention as a substitute for silk in hosiery, nylon in its several different forms has since found widespread use for such common applications as slide fasteners, brush bristles, fishing lines and gears. Nylon films are also frequently encountered in packaging, either by themselves or as laminates with other materials such as aluminum foil or cellulose acetate. Another commonly recognized name among the plastics is polyethylene. Its many packaging applications include squeeze bottles and tubes for cosmetics, toothpaste, and shampoo. Like nylon, polyethylene in film form is often laminated with foil or other films for special applications.

In contrast to the well-known names, other plastics have names familiar only to the plastics specialist. New ones are being developed constantly with an ever-growing range of useful characteristics. Some are soft and flexible, others are tough and rigid. From transparent to opaque, from water-white to brilliant colors, from delicate filaments to strong pipe, from light foam to heavy sheets; these are the property ranges of the plastics that have spurred the growth of the industry from essentially nothing forty years ago to the billion-dollar place it occupies today.

WHAT IS PACKAGING?

The difficulty in defining packaging is a direct result of its many roles in the commercial cycle. All successful packages must: (1) act as a physical container, (2) protect and maintain product quality and (3) appeal and attract the consumer. In essence packaging is a discrete part of the production process. Product economics must include all operations necessary before the product reaches the consumer. The thoughts of all departments in a firm must be integrated into a final package. Users must be made aware of the fact that a manufacturer's main aim is not to sell a package. It is considered a necessary evil and an overhead in selling the product marketed by the manufacturer. Overpackaging is to be avoided in the same manner as underpackaging. Good packaging should be consistent with the demands of the product and be as inexpensive as possible. To insure good packaging science comes into play. Analytical techniques, shipping tests, quality control, and additional factors form the nucleus of good packaging. In the United States (in contrast to Europe) packaging material users look toward converters for new concepts and ideas. Seminars, meetings, and informal discussions are held by users in an effort to elicit new ideas capable of commerciality.

TABLE 1.1

PROGRESS IN PLASTICS DEVELOPMENT WITH
APPROXIMATE DATES COVERING INTRODUCTION OF SOME COMMERCIAL PLASTICS

Year	Plastics	Typical Application
1870	Nitrates (Celluloid)	Billiard balls, eyeglass frames
1909	Phenolics	Telephone hand set
1909	Cold molded	Electric heater parts
1919	Casein	Knitting needles
1919	Vinyl acetates	Adhesives
1926	Alkyds	Molded electrical bases
1926	Aniline-formaldehyde	Terminal boards
1927	Cellulose acetate	Molded products
1928	Ureas	Lighting fixtures
1931	Acrylics	Brush backs, displays
1935	Ethyl Cellulose	Flashlight cases
1936	Polyvinyl chloride	Raincoats
1938	Polyvinyl acetals	Safety glass interlayer
1938	Polyvinyl butyral	Safety glass
1938	Polystyrene	Housewares

Year	Material	Application
1938	Cellulose acetate butyrate	Extended trim
1938	Polyamides (nylon)	Fibers
1939	Polyamide molding powders	Gears
1939	Melamines	Tableware
1939	Polyvinylidene chloride (saran)	Auto seat covers
1942	Allyl diglycol carbonate (CR-39)	Cast sheets
1942	Polyethylene	Squeeze bottles
1942	Polyesters	Laminated reinforced plastic boat
1943	Silicones	Motor insulation
1943	Polytetrafluoroethylene (Teflon)	Gaskets
1945	Cellulose propionate	Pen casings
1947	Vinyl organosols and plastisols	coatings, foams
1947	Epoxies	Potting compounds, adhesives
1948	Acrylonitrile-butadiene-styrene (ABS)	Simulated leather for luggage, etc.
1948	Polychlorotrifluoroethylene (Kel-F)	Gaskets and valve seats
1953	Polyurethanes	Sheets and foams
1955	Polyurethanes	Coatings
1958	Polyacrylamides	Adhesives
1958	Polyethylene Oxide (Radel)	Packaging
1958	Polyacetals (Delrin)	Automotive parts
1959	Chlorinated polyether (Penton)	Pump parts
1959	Polycarbonate (Lexan)	Housings
1959	Polypropylene	Luggage
1962	Polyallomers	Molded hinges

Courtesy of Max Factor

FIG. 1.3. CONSUMER APPEAL

These units display extremely imaginative packaging designed to appeal and attract the consumer.

Packaging Investigates Four Areas

In order to successfully introduce a new package it is essential to investigate four basic areas: (1) material specification, (2) product compatability, (3) distribution cycle and (4) marketing considerations.

Material Specification.—A functional material must have the necessary strength to withstand all the conditions of processing, storage, and transport. If a food product is packaged the material must be devoid of product reaction and not contaminate the food. The material should be capable of preventing ingredient permeation from the inside and atmospheric spoilage from the outside. It is essential that the material be easily sealed on conventional filling and packaging machinery.

Product Compatability.—An investigation of the product should entail its chemistry and physical characteristics. Is it a liquid or solid? Emulsion or solution? Heat sensitive or insensitive? Acidic or basic? The relationship between product and package can best be illustrated by two specific examples.

A packaging system for fresh meat is much different than for cured meat. The biological characteristics inherent in both meats must be considered. Fresh meat requires oxygen in order to maintain red meat bloom. Cured meats survive in a vacuum pack due to different pigments. Clear, unsupported polyvinylchloride film per-

Courtesy of Packaging Science and Engineering,
Rutgers, the State University of New Jersey

FIG. 1.4. THE RUTGERS PACKAGING SCIENCE AND ENGI-
NEERING LABORATORY HAS A WELL-EQUIPPED AND
ACTIVE PROGRAM IN PERMEATION AND LEAK TESTING OF
PACKAGES AND MATERIALS— THE MoCoN PERMATRAN C
ILLUSTRATED IS BEING USED TO EVALUATE THE CO_2
BARRIER PROPERTIES OF MATERIALS AND PACKAGES

forms excellently in retaining red meat bloom. As a cured meat
wrapper it will not yield the long-term shelf-life characterized by
nylon/PVDC/PE (polyamide/polyvinylidene chloride/polyethylene).

In packaging dehydrated foods moisture protection is essential.
The products are extremely hygroscopic. Since most present-day
dehydrated food markets revolve around the military, lightweight
flexible packages are of special significance. Paper/PE (polyethylene)
laminations do not provide the MVTR (Moisture Vapor Transmission
Rate) deemed mandatory for dehydrated foods. Yet, sugar is
successfully packaged in Paper/PE. The increased moisture protection needed for dehydrated foods is only met by the integration
of aluminum foil into a composite structure.

Distribution Cycle.—Package distribution and overall shipping
cycles are perhaps the most difficult part of a comprehensive
packaging evaluation. Uncontrollable variables are rampant and
present-day laboratory tests are deficient. A successful package
should be investigated from the time of its initial packaging to the
time it reaches the consumer. Loading, vehicle movement, ware-
housing and climatic hazards, and merchandising are all variables in a
functional unit. Careful attention must be given to the method of

Courtesy of Packaging Science and Engineering,
Rutgers, the State University of New Jersey

FIG. 1.5. RESEARCH AND TESTING OF PRODUCT/PACKAGE RE-
SPONSE TO THE TRANSPORTATION ENVIRONMENT ARE CRITI-
CAL FACTORS IN IMPROVED PRODUCT PROTECTION

The versatile equipment illustrated is part of the Packaging Science and
Engineering Program of Rutgers University. This MTS vibration tester is
used to evaluate packages and products under a controlled repetitive shock
(vibration) environment.

packing for distribution. Are staples used instead of gummed tape?
Final package weight is also an often overlooked factor. Most
successful packages for retail distribution are in the 30 to 60 lb
range. A lighter package may be thrown carelessly; a heavier unit
may give rise to excessive dropping. Loading on interplant transportation is a critical operation. Packages should be stacked in a
proper manner dependent on the overall geometry.

The method of shipping may affect the final package. In rail
transport three hazards exist: (1) Shunting shocks are due to the
method of train assembly. If very sensitive packages are to be

shipped by train special cans may be necessary. (2) Vibration is due to speed and type of track. Proper package design can usually eliminate vibration hazards. (3) Short starting and stopping trains are an additional danger. Close-coupled trains can eliminate excessive stopping. Truck shipment entails vibration and bouncing dangers. In some cases the use of an "Impact-o-Graph" can offer a method of measuring rough road surfaces. The instrument yields an objective value to all shocks and vibrations encountered on a journey. Air cargo is a rapidly growing industry. Lightweight materials are particularly desirable and entire products are sometimes wrapped in shrink film. An interesting concept exists in packaging fresh strawberries for air shipment to Europe. Cartons of berries are stacked on pallets and polyethylene foam sleeves are placed over each pallet. Ease in handling is increased and moisture retention results. In all types of shipment the concept of containerization is a coming reality. A container is a device used to transport materials in units too large for manual handling. An example of the successful application of containerization is export packing. A highway van is loaded at a shipper's dock without any distinct export protection. At the point of embarkation the container is directly transferred to the ship. Pilferage is eliminated, package handling is minimized and damage is reduced.

Marketing Considerations.—The success or failure of a product is dependent on whether the package is able to sell the product. The shape of the package, graphics, materials and advertising directly influence its success. Marketing is the overall strategy which moves goods from the source of production into the hands of the consumer. Regional tastes and preferences must be considered in all marketing plans. In most Latin American nations prepackaged foods sell poorly since housewives prefer to cook their own meals. Several years ago the introduction of a green colored wrapper into the bread market met dismal failure, yet, a brown wrapper was successful. Shoppers related green to mold and brown to oven-baked freshness. In the Far East white signifies mourning, while purple is related to death in Latin America. Hot, strong colors sell excellently in tropical lands and blues, gray and greens are successful in cooler climates. An interesting discovery by package designers was the failure of blue packages in the Hong Kong market. Chinese consider a blue color unlucky while red is related to prosperity. Color also connotes a product image. Eastman Kodak yellow is a classic example. Strict quality control is maintained in order to insure adequate color reproducibility in all Kodak products.

Economic considerations are of major importance. In a low-income area smaller units sell better than the larger economy size. Even though a large package offers price-savings many low-income shoppers can only affort to purchase a small package.

European packaging concepts differ considerably from those of the United States. Wax laminated cellophane is popular in Europe. In the United States most consumers believe it feels greasy. In Europe many condiments are sold in flexible tubes. Yet, in the United States foods in tubes are related to toothpastes. A challenge to the packaging fraternity is to provide a standardized design capable of crossing national and international boundaries. It is an extremely difficult result to accomplish since there is a tendency toward demanding new experiences and excitement. The converting industry is dependent on everchanging packages. Marketing and advertising are invaluable assets to creating markets by stimulating demand.

Packaging Education

The rapid growth of packaging technology has been one of the characteristics of modern society. In the United States consumer affluence has given rise to newer foods and the demand for newer packages. This trend has carried over to Europe and supermarkets are now part of the European scene. Although food production and packaging has developed in the advanced countries, the statistics for food production in the developing nations are rather startling. More than half of the world's population lives in the Far East. However, only 19% of the world production of foods of animal origin and 44% of the production of vegetable foods are processed there.

Packaging has developed, but the need for packaging education has been sorely neglected. Only within the last few years has packaging been recognized as a separate discipline in the more sophisticated nations. In lands where supermarkets do not exist packaging means more food, and not merely a show of national affluence. By applying packaging principles olive spoilage in Chile was reduced by over 60%. Other cases are even more dramatic. It is essential that the developed lands export packaging technology to the rest of the world.

Few Texts Available.—Since the end of World War II, not more than a dozen books have appeared discussing the basic principles of packaging. Two excellent texts have been published by The Institute of Packaging (U. K.), *Fundamentals of Packaging* and *Packaging Materials and Containers*. One of the prime movers in the need for packaging education has been the European Packaging Federation. Their *Blueprint for Packaging Education*—1967 is a classic publication. An outline for packaging education is described in workable and detailed terminology. Perhaps the first text devoted to the

Courtesy of American Spice Trade Assoc.

FIG. 1.6. PICKING CLOVES

Proper packaging will insure the ultimate profitable marketing of this product around the world.

protective packaging of various products, including food, appeared in 1957 (second edition in 1965). Written by Oswin and Preston of British Cellophane, Ltd., it admirably covers the entire industry in severely limited space. *Economics of Packaging* by Edmund Leonard is the first book ever published discussing the cost factors inherent in good packaging. A more recent text devoted to the packaging engineer has been J. R. Hanlon's *Handbook of Package Engineering*. This author in conjunction with R. C. Griffin, Jr. has written four books for packaging personnel.

Leadership Has Been Lacking.—In the United States students involved in both undergraduate and graduate degrees in packaging have few texts. Various journals and compilations of information are used, but these are not truly academic. Professors utilize notes and distribute these to their students. Even in food technology curriculums courses in food packaging are the exception rather than the rule. Why? Simply because no text is available for student study. Yet it is these students that will eventually be responsible for developing new foods and their corresponding packages.

Poor Planning Leads to Self-destruction.—A new industry must plan for the future and insure the constant flow of trained personnel. Packaging has evolved rapidly. Two colleges now offer an undergraduate degree in packaging, and two universities have graduate programs. But these are not enough. Public recognition for a new industry can only come about when a degree in packaging becomes as commonplace as one in chemistry or physics. If the packaging industry does not properly train its personnel severe results can occur. To draw an analogy, in 1930 there was a plentiful supply of garment workers in the United States. The garment industry was strong and productive. Yet in 1975 there was a severe shortage of trained garment workers. Older people have died and no one has replaced them. In the most productive years no leadership appeared to spearhead the need for training. Schools were virtually nonexistent and management talent was not exploited. The result is the current shortage in trained workers.

Food Packaging Courses Should Be More Commonplace.—The logical place to start with a more sophisticated approach of packaging technology is the food industry. Food is essential to human survival. An extremely large proportion of the packaging industry is involved in food packaging. An interesting approach has recently been introduced by the Packaging Association of Canada. This organization has supplied a course on tape and slides to the FAO International Food Technology Training Centre at Mysore, India. Based on its own success, the courses are being used for both undergraduate and graduate studies. In addition the FAO has sponsored mobile training units which visit food plants and offer guidance in packaging technology.

A more conventional and practical approach in the United States is simply to offer more food packaging courses at the university level. Every student involved in food technology should be given a 12-month course in packaging for degree credit, and students involved in nutrition, biochemistry and the more peripheral fields a 6-month course. These courses must not be elective but required ones. It is the package that contains, preserves, and sells the food. A food without a package becomes an academic curiosity, not a commercial success.

Results of Education May Decrease Waste.—Statistics describing the actual loss between food production and consumption are rare.

For certain fruits and vegetables more than 50% loss is estimated. An overall value of 50% loss has been given for all foods produced in the more advanced nations. Estimates of loss of canned foods stored in the tropics and subtropics reach 11%. More than 25% of all dried fish is lost due to spoilage during storage. Coupled with possible losses in nutritional values, the differences between produced and consumed foods become significant. If losses can be reduced by only one-half many nations can become self-sufficient. Packaging works in direct opposition to waste. It can help in eliminating hunger in the world. And eliminating waste is just as important as creating new foods. Education in packaging must be recognized by the packaging industry. It is immaterial whether the motive is one of survival or the elimination of hunger. Packaging education should not be neglected. It is vitally needed.

Courtesy of Romanoff Caviar Co.

FIG. 1.7. FISHING FOR STURGEON

Workers at the Iranian Fisheries in the Caspian Sea bring in their harvest of sturgeon before removal of the precious roe. If the caviar is not packaged for export, a valuable commodity is lost.

BIBLIOGRAPHY

ADLER, E. 1974. Consumer attitudes to packaging. Packaging News (Suppl.) 9–10.
ANON. 1973A. Choosing alternatives in a tight market. Mater. Handling Eng., Spec. Issue, Fall, pp. 32–36.
ANON. 1973B. Important steps to eliminate packaging equipment downtime. Can. Packaging 26, No. 12, 20–23.
ANON. 1974A. Cracking the logo codes on packages. Packaging News (Suppl.) 6–7.

ANON. 1974B. Helping the retailer reduce in-store losses. Mod Packaging *47*, No. 1, 36-37.

ANON. 1974C. Low cost packaging for protection and easy safe movement. Packaging Rev. *94*, No. 4, 23,25-26,29.

ANON. 1974D. Coping with shortages. Mod. Packaging, *47*, No. 4, 26-28.

ANON. 1974E. Learning to live with UPC symbol. Mod. Packaging *47*, No. 5, 20-22.

ANON. 1974F. Facts from figures—packaging statistics unified. Packaging News (Suppl.) 16.

ANON. 1975A. Packaging applications and trends. Food Manuf. No. 9, 45-50,52,54,58.

ANON. 1975B. Study cites energy reductions. Food Drug Packaging No. 4, 4, 10.

CHADSEY, R. 1974. In defense of packaging. Can. Inst. Food Sci. Technol. J. No. 12, 749.

FARRAWELL, M. 1973. The role of quality control in packaging. Australian Packaging *21*, No. 9, 55,57,59-61.

GOOCH, J., and PAINE, F. 1973. Trends in packaging. Paper, pp. 108-109,111,120.

HAUBER, W. 1974. Packaging in '74: The need for innovation. Flexo. Packaging Printing *19*, No. 3, 24-25.

HUTT, P. B. 1973. Safety regulations in the real world. Food Cosmet. Toxicol. *11*, No. 5, 877-884.

MacCHESNEY, J. C. 1974. Packaging men must grasp the supply nettle now. Packaging Technol. *20*, No. 134, 4.

MAGRAM, S. H. 1974. Labelling regulations. Aerosol Age *19*, No. 1, 28-30,38.

NARRACOTT, E. S. 1974. Plastics for conservation. Plastics Rubber Weekly No. 524, 7.

NEUBAUER, R. G. 1973. Packaging. The Contemporary Media. Van Nostrand Reinhold, New York.

OPATOW, L. 1973. What do consumers want from packs today? Imballagio *24*, No. 212, 27-28. (Italian)

PAINE, F. A. 1975. What packaging means to the quality of life. Packaging Technol. No. 13, 1-4,16.

PESSOL, P. R. S. 1973. Fundamental needs and methods of mail order packaging. Packaging Technol. No. 132, 13-15,30.

SEYMOUR, R. B. 1975. Packaging: The American angle. Plastics Rubber Weekly No. 567, 16-17.

SIMMS, W. C. 1974. Ecology ratings—a new frontier. Mod. Packaging No. 7, 18-21.

TAKATSUKI, K. and HAYASHIBARA, I. 1973. Quality control of packaging materials. Japan. Plastics Age *11*, No. 12, 23-28.

CHAPTER 2

Glass Packaging and Closures

INTRODUCTION

The origins of glass are lost in antiquity. Thought to be an Eastern Mediterranean discovery of about 3000 B.C., it was known to the predynastic Egyptians. Early Egyptian glass was used in the form of a glaze for stone beads. Some of the earliest small vases were probably made by dipping or rolling a clay or sand core in molten glass. As an additional later step, softened rods of glass were wound around a core. While several historians claim that this technique was Syrian, it was the early Egyptian glassmakers that learned how to select the best types of silica and the most efficient fluxes. The early "sand-core" method was eventually displaced by the pressing of bowl shapes. Sometime in the 1st century B.C. glassblowing appeared in Syria. Glassblowing was a much used technique in the Phoenician trading city of Sidon. The art was a well-developed method used by the ancient Hebrews and it was after the Diaspora that these same Hebrews introduced glassmaking to Byzantium.

The Roman glassmakers learned their techniques from both Syrian and Egyptian (Alexandria) emigrants. Glassmaking flourished under Imperial Roman rule. While the Romans were not at first makers of the finest decorative glass, they did have a peculiar genius for establishing factories for quantity production. Glass was widely used in ancient Rome as walltiles and as mosaics. Common domestic Roman glass was usually a square bottle blown into a mold with the handle added in a subsequent step. Under the Romans great centers of glassmaking were established. When the Romans lost power these centers gradually declined in importance. Stained glass was an important European manufacturing technique in the Middle Ages. Great cathedrals loomed with magnificent stained glass windows.

In the Orient blownglass was probably made during the reign of Emperor K'ang Hsi (1662-1722). It is also almost certain that the introduction of the secret of glassmaking into China came from Near Eastern sources. With the rise of Islam in the 7th century A.D. glassmakers spread to Persia and parts of North Africa.

European glassmaking revived first in Italy during the 10th century. Fine glassmaking flourished in Venice and by the 15th to 16th centuries Venice was the leading source of fine glass.

In 1608 glassmaking was introduced to North America when the London Company of Virginia started a small factory in Jamestown.

17

None of the production has survived, however, it is known that they made mostly bottles and window glass. In 1739 Caspar Wistar (1696–1752) started the first successful glass factory in southern New Jersey, producing greenish bottle and window glass. Another pioneer in early American glassmaking was Henry William Stiegel (1729–1785) of Pennsylvania, who made fine tableware. One of the first bottle-making machines was invented in 1882 by Phillip Arbogart of Pittsburgh. His technique was based on a press and blow method. Closures began to appear in the early 20th century. The roll-on closure was first developed in 1923 while the pry-off cap appeared in 1926.

Early Uses of Glass in Packaging

Used to contain cosmetics, perfumes and foods by the ancient Egyptians, Hebrews, Romans and Persians, glass was also first used centuries later by Appert of France for food preservation. The limited supply of glass bottles severely hampered their use in packaging until the late 1890's. By that time John Mason had already invented his screw-topped jar and Louis Pasteur had discovered his methods for food preservation by the destruction of micro-organisms. He used glass containers. Glass was widely used as medicine bottles in the late 19th century. Appearing as food packages in the 20th century, glass bottles have played a truly vital part in the growth of the packaging industry. They are produced at low cost and within small dimensional tolerances. They also are capable of being run through filling lines at very high speed.

Glass Composition

Glass is a substance that is hard, brittle and usually transparent. Soda-lime glass is made from limestone (about 10%), soda (about 15%) and silica (about 75%). Lesser percentages of aluminum, potassium and magnesium oxides may be included. When the components are melted together they fuse into a clear glass which can be readily shaped while in a semimolten state. The melting of glass is usually accomplished at temperatures of about 2800° F in a very large furnace (usually gas heated). These furnaces are often described in terms of their surface area (i.e., 800 sq ft, with a depth of about 3 to 4 ft) and constructed of heat-resistant refractory materials. Because of the high amount of heat required to bring such a large mass of material up to 2800° F, these furnaces are run continuously except for maintenance shutdowns. The life of the furnace walls is about three years. Good glassmaking depends on maintaining steady-state conditions. Variations in formulation can provide differences in glass properties. Amber-colored glass can be

Courtesy of Owens-Illinois, Inc.

FIG. 2.1. GLASS PACKAGES IN THE MARKETPLACE

Here is a representative sample of the glass bottles and jars produced by Owens-Illinois, Inc., the world's largest manufacturer of glass containers. On the right is the relatively new Plasti-shield® container, a lightweight convenience glass bottle wrapped in a foam polystyrene sleeve which keeps the beverage cool longer. The sleeve is predecorated and saves the bottler from having to label each container.

produced by adding iron oxide to the melt. For yellow-colored glass cadmium compounds are usually added. Opal-colored glass involves the addition of fluorides or phosphates. Often broken glass or "cullet" may be returned to the melt furnace. If added in significant quantities it may alter and/or color the resultant glass. Additives result in a large tonnage of glass melt which must be converted into glassware. Unless the packager is a very large user, he always must accept the basic glass formulation offered by the glassmaker. The latter cannot afford to convert an entire batch to produce say 10,000 bottles of a special glass.

There are many other types of glass which are produced for special applications. Although not usually in use for packages (soda-lime glass is the most widely used) they are of interest to the reader.

Lead glass is made from lead monoxide with sand, sal soda and small amounts of other materials. Lead glass is commonly used to make lenses, art objects and fine tableware. Additional uses include gamma-radiation-shielding windows and television tubes.

Borosilicate glass is made by mixing silicon dioxide, boric oxide and aluminum oxide. This type of glass does resist heat, shock and most types of chemical damage. Commonly used for ovenware, laboratory equipment and high-voltage insulators, it is also a recognized type of glass in the USP for containers.

Silica glass is made with the same ingredients as borosilicate glass. The mixture is first treated with acid, leaving a porous substance. It is heated to a high temperature and the glass shrinks. The pores close and a transparent, nonporous glass is the result. Due to its great tolerance to sudden temperature changes, silica glass is often used for ovenware, sun lamps and other industrial applications.

Photosensitive glass is that which produces images from photographic negatives. The image emerges after the glass is exposed to ultraviolet light, followed by heat, chemical or other treatment.

Other types of glass that are manufactured include electrical-conducting glass, fibrous glass and cellular glass.

Courtesy of Farbwerke Hoechst AG.

FIG. 2.2. INSULIN PRODUCTION

Manufacture of Insulin is a complex process. It requires a USP glass for proper protection.

In drug packaging four types of glass are recognized in the USP for container application: (1) Type I neutral—borosilicate glass; (2) Type II—soda glass with a surface treatment; (3) Type III—soda glass of limited alkalinity; and (4) NP—soda glass (for nonparenteral use).

Glass Properties

Because of its outstanding properties glass remains a most important material in the packaging industry.

These major properties are: (1) chemical inertness, (2) non-permeability, (3) strength, (4) resistance to high internal pressure, (5) optical properties, and (6) surface smoothness. On the negative side, glass containers are also fairly heavy and yield a higher tare than most other packaging materials. They also are generally quite fragile.

Chemical Inertness.—Glass usually exhibits a high degree of resistance to chemical attack. Only hydrofluoric acid (HF) appreciably attacks glass. Since there are many different types of glasses available, some are more chemically resistant than others. Variables involved in this property include chemical composition of the glass, time and temperature of attacking reagent, and whether the glass has been in contact with other hazardous elements. Neutral glass is generally more resistant to chemical attack than alkali glass. The latter can yield alkali to aqueous solutions and thus affect either suspended or dissolved materials.

Non-permeability.—Glass is completely impermeable to all gases, solutions or solvents. In most respects it is the "perfect" packaging material as regards gas transmission. The ability of a glass container to be the ultimate barrier is by far the strongest asset to glass as a packaging material.

Strength.—The strength of glass is dependent on the type of glass involved. It is the condition of the surface that influences its tensile properties. When glass fails it is a result of tensile stresses and never stresses due to compression. The theoretical breaking strength of glass is over 100,000 kp per cm^2. The general practical strength of glass containers is in the order of 200 to 1,000 kp per cm^2. Less than 1% of the theoretical strength is utilized. Tempered glass has greater strength and is used for automobile windows and other specialized applications.

It is quite important to note that a smooth surface on glass results in a fairly high strength value. When the surface becomes scratched or bruised, it results in a drop of the strength value to a fraction of the original strength. As much as half the original strength may be lost as a result of scratching. This strength can be restored by etching the surface with hydrofluoric acid, which removes a layer of glass only a fraction of a micron thick.

The big difference between the theoretical strength of glass and its practical strength is the various imperfections at the surface. These imperfections act as stress concentrators. Failure of the glass results from the propagation of one of these flaws. Glass fractures always start at the surface and propagate outward from it. The fracture pattern and the fracture surface will often exhibit certain features. A container may break in single or multiple stages. There may be inside surface damage, impact star-crack, internal pressure completing the breakage, or a butterfly bruise in the heel region and thermal shock completing the breakage. Secondary breakage may also occur without any direct relationship to the actual fracture process. An example of this would be a mineral water bottle breaking primarily from internal pressure and the various glass pieces subsequently being broken by impact when they are thrown against a concrete floor.

Various coatings can be sprayed on glass in order to preserve its original strength. Often they act as lubricants so that contact between bottles is less likely to cause damage. Types of coatings used include silicone, wax and various resins.

Resistance to High Internal Pressure.—Due to the very high rigidity and strength glass is able to withstand high internal and/or external pressures. Its heat resistance and high melting point enable glass containers to be used as packages for both moist and dry sterilization. While borosilicate glass is particularly good for retarding thermal shock, soda glass is quite suitable for hot filling operations.

Optical Properties.—Glass has a high degree of transparency when clear. When opaque or amber-colored containers are used light

TABLE 2.1
VITAMIN C OF PACKAGED MILK BEFORE AND AFTER EXPOSURE TO DAYLIGHT

Exposure Time (Min)	Glass (mg/l)	Glass (%)	White PE Film (mg/l)	White PE Film (%)	White/Black PE Film (mg/l)	White/Black PE Film (%)
0	13.5	100	13.5	100	13.5	100
55	1.3	9.6	2.1	15.5	11.8	88
120	0.6	4.3	1.0	7.4	11.4	84
240	0.0	0.0	0.0	0.0	11.5	85

Source: Briston (1968).

deteriorative effects are significantly reduced. The degree of protection will depend upon the extent to which glass excludes light of the wavelengths to which the contents are sensitive. Amber glass used commercially for blown bottles (2 mm thick) excludes practically all light of wavelengths lower than 4500 Å. The corresponding figure for

TABLE 2.2
RIBOFLAVIN LOSS IN MILK AFTER 1¼ HOURS IN VARIOUS PACKAGES

Packaging Medium	Loss of Riboflavin (%)
Glass	32.4
PE film (transparent)	39.5
PE film (white pigmented)	30.4
PE film (black/white)	2.7

Source: Briston (1968).

green glass is about 3500 Å. As the thickness and glass distribution vary, light filtration is affected.

Surface Smoothness.—In a clear or polished glass, there exists a high degree of brilliance or sparkle which is often considered an asset in sales appeal. Cleaning is made easy by the smooth surface of glass.

Glass Container Design

Glass Surface.—The condition of the glass container as received by the packer is extremely critical in its ultimate successful utilization and proper handling. Containers should arrive at the plant in as good a condition as possible. The theoretical strength of glass is over 100,000 kp per cm². Glass containers should arrive for use at 800 to 1,000 kp per cm². When the glassware leaves the annealing lehr, its surface is extremely clean and is quite sensitive to scratching. If glassware is allowed to cool too quickly strains are set up and the container becomes more susceptible to damage. In the lehr the temperature is raised to 1000° F and held for 15 min. The containers are then cooled very slowly to room temperature. Strains (known as cords) may occur. Caused by poor mixing in the melting tank they produce weak containers. Damage received by the containers at the end of the annealing lehr is often the most important factor in determining the strength of the container. Visual inspection and crating can inflict damage to a sensitive container's surface. The use of coatings on the surface will often alleviate the entire problem.

Container Construction.—How to design a useful glass container? There is no easy answer to this extremely difficult problem. Should the marketing or technical aspects be considered as a prime factor? Shape is a critical factor. A sphere is the strongest shape, while cylindrical and rectangular shapes are poorer. In the design consideration is given to directing the shocks to the area of the bottle where they will do the least harm. Ridges or bands may be designed at the shoulder and base of the container to absorb any impact stresses. These ridges are knurled rather than straight. Stippling of the base is also useful in improving base grip and container strength.

Courtesy of Owens-Illinois, Inc.

FIG. 2.3. GLASS FORMING

This is the forming area of the glass container manufacturing plant of Owens-Illinois, Inc. Melted glass is formed into glass bottles and jars on 4 high-productivity glass-forming machines which operate 24 hr a day, 7 days a week. Although the machines are automatic, they are constantly checked by skilled operators.

A considerable amount of damage to the lowest 15 to 25 mm of the side wall can be avoided by undercutting the lower part of the wall, making it more difficult for this point to be hit. An additional source of damage is the chipping of the inside surface at the center of the bottom panel. If containers holding heavy syrups are dropped they may develop a vacuum bubble at the bottom. This bubble forms and collapses rather quickly, but it may break the container surface and cause subsequent weakening.

Glass Container Fabrication

Glass containers may be made by the following methods: (1) extrusion, cutting and shaping; (2) press and blow, or blow and blow (blown); and (3) pressing.

The raw materials used in glass manufacture are sand, soda, ash, limestone and cullet. These products are measured and are then mechanically mixed into a batch. The molten glass is about 4 ft deep and is maintained at about 2700° F by gas flames across the top surface. From the furnace molten glass flows in a steady stream through a narrow throat and then into a feeder channel. The molten glass is gathered through an opening and chopped off by a pair of shears into "gobs" which fall into blank molds. The amount of glass obtained depends on the temperature, glass composition, orifice size and shear timing. Double-gob blow molding means extruding two gobs at once to increase the speed. Triple-gob machines making three containers instead of one are in use for 12-oz beer bottles. The gobs feed single, double or triple cavities of an individual section blowing machine. The blank mold is the mold in which the first stage of shaping of the container body is performed. In one alternative, the suction process, the finish is at the top and the mold sucks the molten glass from beneath. In another more common flow process the molten glass flows downward into a blank mold with a reciprocating plunger assist.

Air is blown on the molds to cool the glass and the molded glass is removed from the mold where further cooling occurs. When the glass leaves the mold the surface is rigid, but the core is still hot and soft. Were natural cooling to continue, hot inner parts could contract more than the cool outer surface, and stresses would be established with the inner parts stretched and the outer surface compressed. To avoid these dangerous differential stresses the glass is annealed in a tempering oven called a lehr. Glass is conveyed to the lehr where it is reheated to a temperature at which the glass can flow slightly and relieve the internal stresses, and then cool in a uniform manner to avoid any further stresses. Annealing requires over 1 hr of controlled temperature. This process could be performed in the mold, but only by tying up the mold for the requisite time period. During annealing the finish (so called because in the days of manual glassblowing it was the last operation) is fire-polished to smooth any imperfections. In the blow and blow option of the flow process, the finish (the molded portion to which the closure is applied) is at the bottom of the cavity and is formed. The pattern or parison is rotated and transferred into a finishing or blow mold where compressed air blows the parison into the shape of the mold. The blow and blow process is usually used for the narrow neck or bottle shapes.

In the press and blow process the gob flows into the blank mold. The parison is pressed to an exact shape by a plunger before it is transferred to a blow mold where the final shape is blown. Press and blow is used for widemouthed containers or jars.

Molds are a costly investment—about $3000 to $8000 a set depending on the size and complexity. A set of molds lasts about 2.5 million bottles. Glass industry practice is that the packager directly pays for the first set of molds (even if not used up) and the glassmaker assumes the direct cost of subsequent sets (although, of

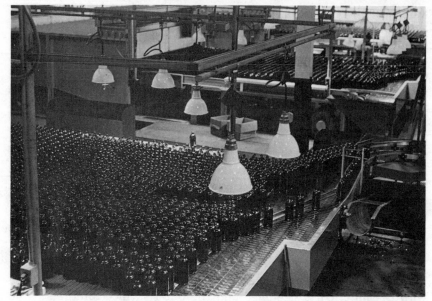

Courtesy of Owens-Illinois, Inc.

FIG. 2.4. END OF MANUFACTURE

This is the cool end of the annealing oven. Each container is inspected several times prior to being shipped to customers.

course, the packager pays in variable costs). Blank and blow molds are machined to precise dimensions usually from fine-grained cast iron.

Theoretically, very close inspection is given to the glass to ensure its dimensions, roundness, verticality, volume, weight of glass per container, etc. Reports by the National Commission on Product Safety would indicate that the level of quality testing, although perhaps satisfactory for the glassmaker or even the packager, might not be adequate for the consumer of glass-packed carbonated beverages. The Commission has suggested that in the interests of consumer safety each container be individually tested to minimize the explosion hazard.

Costs of glass packages generally increase with weight of glass per package and (although mechanical strength increases with glass thickness) an economic balance must be attained. Glass may be shaped to relatively precise dimensions to control quality of contents by height of fill. A total glass package is not simply the glass container itself, but also a closure and an identification. The latter may be fired on the glass after molding or may be affixed as a paper (foil or film) label. The cost of a permanent decoration is individually far higher than an adhered label, but this differential vanishes when used on a multitrip container. A disadvantage of the permanence of a fixed decoration is that it could make the container obsolete before its shape or closure is out-of-date.

Closures—an Overview with Glass

The type of product, application and consumer use dictate the closure, which in turn dictates the finish. Among the closures that would conventionally be designed are:

Type	Typical examples
plug	wine, liquor
crown	beer, soft drinks
twist-off crown	beer, soft drinks
roll-on	beer, soft drinks
continuous thread	instant coffee
lug or interrupted thread	jams, jellies
press-on	baby food

Originally crowns were intended for single use, but the development of the inexpensive aluminum roll-on for reclosure led to the twist crown (and to an upsurge in the sales of larger bottles that can be opened and reclosed after withdrawal of only part of the contents). Plugs are almost traditional with wine and liquors and so are not being changed too rapidly. Jams and jellies, being reusable items that are packed under vacuum, require a screw-type closure. Instant coffee is not preserved with vacuum but it still requires reclosure. Baby foods require limited reclosure, vacuum, and resistance to high temperatures as well as very high-speed application. The press-on with plastisol interior side walls that can accept a groove impression has proven satisfactory. Closures also may be decorated. Metal lithography is the most widely used method. Molded plastic is used for many nonfood applications. However, the high cost of molds for injection molding deters many food packagers from the investment, especially when metal closures are widely available at low cost.

Because of the huge investment required in equipment glass package making is not an in-line operation. Glass packages are completely fabricated elsewhere and then sent to the packager. In anticipation of orders some glassmakers produce both stock and custom packages for inventory and then draw from these inventories. For some stock containers a single run per year is the only economic quantity. Thus the packager could conceivably find a package that fits the requirements from among the many stock designs. But the capacity and finish (which dictates to a major degree the closure) would be fixed.

Beer and carbonated beverages are in glass because of the flavor inertness, and resistance to internal pressure and compressive strength. (Both beverages are produced in such prodigious quantities that warehouse stacking is mandatory.) Further, neither product undergoes severe heating in processing. In the early days glass was the only packaging material that could be realistically recycled without significant loss of properties (and it may still be).

Glass is also used for pharmaceuticals and other health aids because of its inertness and image of cleanliness and purity. Many products are in dry form and would not react, thus, several have been switched to other less expensive packages. Those products that require extensive shelf-life without change of potency are often glass-packed to protect against evaporation or moisture. The extensive use of glass for toiletry and cosmetic products is due to the ability of glass to retain highly desirable aromatics and perfumes. Since a major attribute of these products is the fragrance, any loss or other change is undesirable. The glass container not only prevents loss by volatilization, but does not allow access of oxygen, interaction of product essential oils with the package, or access of aromas from external sources. Coupled with an effective closure, a glass package can retain quality of fragranced cosmetic toiletry for many years. In consumer use the glass package can be reclosed to retain the fragrance.

Glass packs are used to a considerable extent in the food industry. Probably the broadest usage is for baby foods which converted from metal cans almost completely only a few years ago. Except for baby foods and asparagus spears, very few low-acid food products requiring retorting are packed in glass. In order to ensure that the closures stay on the jars during the cool down cycles of autoclaving, controlled counterpressure must be applied, and this is a complicating factor. Further, retort loading and unloading requires care to prevent breakage. Unless there is some marked benefit, such as showing intact asparagus spears or whole sausages, few food processors slow their cycles to employ glass for products that are more easily processed in other package forms. High-acid foods such as fruits, however, can be conveyed through hot water baths with little fear of breakage during sterilization. Light-colored fruits such as peaches discolor, however, and are rather poor in appearance. Some juices are glass-packed. Large quantities of pickles are glass-packed, but these require only very mild heat processing. The strong aroma of pickles should be kept away from other products.

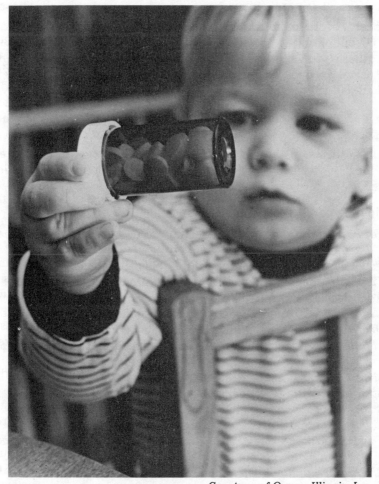

Courtesy of Owens-Illinois, Inc.

FIG. 2.5. YOUNGSTERS FIND THE NEW SCREW-LOC VIAL EXTREMELY
DIFFICULT TO OPEN, SAY OWEN-ILLINOIS OFFICIALS—THE AMBER
VIAL WITH THE WHITE CAP WAS DESIGNED TO REDUCE ACCIDENTAL
POISONING.

The number of glass packages produced continues to increase each
year. Losses to paper in fluid milk and to metal in beer and
carbonated beverages have been offset by increases in baby foods,
instant coffee, and nonreturnable glass for beverages. Although the
glass industry does not manufacture all the glass packages filled and
distributed, it is claimed that glass containers are probably the
package most widely used in the United States today. The industry
produces and ships over 40 billion units (sold and reported as gross,

making for yet another confounding factor), and the estimates for number of fills for returnable beer, soft drinks, and fluid milk are in the 50 billion class. This estimate is based on the number of crowns and other closures shipped plus the average trippage reported by glass packagers. Perhaps as many as 90-to-100 billion glass packages are packed annually.

CLOSURES[1]

One specific type of closure (known as hermetic) prevents no exchange between the contents and the outside atmosphere. It may be loosely termed "air-tight" (dictionary definition—perfectly closed).

Various Types of Closuring or Sealing

Fold—overwrapping without sealing, fold over, fold down, grocers fold, roll wrap, bread wrap, etc.

Twist wrap—i.e., sweets, special flexible cellophane.

Ties—wire, paper wire, plastic wire, plastic for sacks, liner bags.

Adhesives—glues, self-adhesive (envelopes), heat sensitive, dextrin, P.V.A., hot melt, waxes (microcrystalline).

Heat seals—weld (melt), ultrasonic, high-frequency, impulse.

Taping—Self-adhesive, dextrin (kraft tapes), cloths and laminated tapes (Bitumen, foil, fiber), paper, etc.

Metal—nails, staples, screws, tacks for wood, fiberboard.

Strapping—wire, tensional steel, nylon.

String and rope—and stitching (sacks), hemp, nylon, cotton.

Labels, tags and *seals*

Foil—diaphragms, heat seal (blister packs and injection-molded containers), covers (milk bottles), capsulation (wine bottles).

Shrink wrapping—shrinkable films, polyvinyl chloride (PVC), polypropylene, polyethylene (sleeve or complete overwraps).

Crimping—fold and crimping—Collapsible tubes (metal) use double or triple (saddle back) fold. Seal may be improved by wax, latex heat seal or pressure-sensitive band at fold areas; aerosol valves for glass bottles; overseals for injection vials.

Swaging—to shape with a swage (a die or grooved block for shaping), aerosol valves in metal cans.

Closures for Containers

Steel shipping containers—Narrow neck: press caps, screw caps, pourer taps. Full aperture: wide lug cover with sealing gasket.

---

[1] From material supplied by Mr. Dixie A. Dean, Fisons, Ltd., Pharmaceutical Division (U.K.).

Fiberboard drums—slip lid, wide lug over with sealing gasket.

Large aluminum containers—screw caps, bungs and plastic shives (antibiotics and perfumes).

Collapsible Tubes

Metal—blind end nozzles or metal diaphragm; screw caps: thermoplastic no wad (wadless) or thermosetting (wadded).

Plastic—pliable material: rigid wadless caps thermosetting or H.D. polyethylene, polypropylene, polystyrene. Rigid material: wadded rigid cap or wadless pliable (thermoplastic) cap.

Metal can—(1) General line-type containers: lever lid (paint), lever lid and diaphragm (coffee), slip lid (mustard) may be metal or plastic, vacuum lid with flowed in PVC compound (tobacco), tagger top and cutter slip lid (cigarettes), hinged lid (cigars), screw cap (metal or plastic), rotary (plastic) (talcum powder), Jay-Cap (thermoplastic), crown cap. (2) Open-top cans: ring (pull) tabs on double-seamed aluminum lid (beer), key opening (sardines) on double-seamed food cans (ham), double-seamed ends (compound lined). Note use of five thicknesses of metal (all food-type cans). Note also closure or sealing of other seams may be overlapped and soldered, single seam (dry, soldered, cemented or doped), welded.

Extruded aluminum cans—These are widely used in the United Kingdom for pharmaceutical tablets but are not popular in Europe or the United States. Closures are either plug (polyethylene, snap-on over-bead (polyethylene) or metal screw cap fitted with conventional wadding or flowed-in compound lined.

Flexible Packaging

Laminates, films, pouches, sachets etc.—Seals may be fin (inside to inside) from a single web, overlap (inside to outside) from a single web, inside to inside from two separate webs (four-sided seal), or one web folded over (three-sided seal). Sealing may be achieved by heat or adhesives. The three parameters to consider in sealing are: (1) temperature, (2) dwell and (3) pressure. Heat seal may use direct heat, bonding being affected either by a heat sealable coating (i.e., nitrocellulose, PVDC, etc.) or by the direct welding characteristics of the material used (i.e., polyethylene). The source of heat may also be created by ultrasonic, radio (high) frequency welding, or impulse. Seals can also be achieved in certain cases between two different materials (i.e., melted or softened polyethylene will adhere to paper or board). For

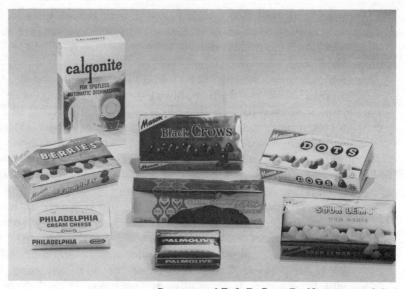

Courtesy of E. I. DuPont De Nemours and Co.

FIG. 2.6. HEAT-SEAL-TYPE CLOSURE

Tighter, more attractive packaging at lower cost is result of heat-sealable foil overwraps using inner hot-melt coating of DuPont's Elvax vinyl resin and paraffin wax. Structure eliminates one layer from conventional foil wraps, saving as much as 8% in cost. High bond strength also offers greater package protection. Commercial uses range from bar soap to candy and cream cheese.

example, plastics suitable for skin packaging include: vinyl, ionomer, polypropylene, acetate and butryate. Although many films, laminates, etc., may have a 100% effective seal, the materials employed may be permeable to moisture, oxygen, carbon dioxide, etc. This property can be either a disadvantage or an advantage. For example, MSAT[2] film overwrap can permit sufficient moisture permeability to cause a biscuit to reach its critical (unpalatable) moisture level.

Paper and board containers—Certain carton systems (Hermatet, Cekatainer) offer closuring systems which give a high protection against moisture/gas exchange. These rely on a heat sealing medium (polyethylene, ethylene vinyl acetate (EVA), ionomer, PVDC, etc.) plus the barrier properties given by those materials. When used in conjunction with aluminum foil, excellent protection can be achieved for both dry and liquid materials. Closures for powder products are

---

[2]Moisture-proof, heat-sealable, anchored, water resistant grade of cellophane made by Olin Corp.

improving but do not yet match the original seal. Other waxed or coated cartons are also in use (i.e., tubs, Tetrapak). The more conventional cartons rely on friction or locking flaps for closure or adhesives. Flip top (as in cigarette packages) is a more recent innovation.

Composite Containers—These have either metal or metal and plastic ends. Closures may be lever lid, slip on (in or over) metal or plastic. As with the special Hermatet type of cartons, product usage now includes both liquids (motor oils) and powders. Foil is again used to give the barrier properties.

Fiberboard outers—Outers may be closed by interlocking flaps, glue, tape, staples or a combination of these. Interlocking flaps are frequently used when bottles, cans, etc. are delivered in "own cartons." Bottom of case is sealed and interlocking flaps are placed downwards (inverted) on pallets to prevent ingress of dirt, etc. Glue may be silicate, dextrin, PVA vinyl acetate copolymer or hot melt (spot or fully glued).

Tapes—gummed tapes (kraft, crepe, cloth, union kraft, etc.), film tapes (cellophane, acetate, vinyl, polyester or film reinforced with rayon, glass or nylon), metal foil tapes (usually aluminum foil to improve moisture barrier properties). Note: H-type sealing gives greatest protection as it will exclude dirt, dust, etc.

Bags—paper, plastic, laminates. Paper bags may be stitched, stapled, sealed with a header label or adhesives, or tied. Plastic bags can be stitched etc., as paper bags, but are more conventionally heat sealed. Laminated bags (same as plastic).

Preformed Screw Closures

Historically screw closures have consisted of a threaded shell containing a wad and a facing. The wad is made from cork agglomerate (gelatin or resin bonded), pulpboard, or feltboard. This provides for a resilient backing into which the seal is "bedded." The wad may be backed with kraft paper and/or hot waxed if required. A facing, which must be resistant to the product, may be one of many materials:

Aluminum foil—0.025 mm or the same with a microcrystalline wax coating.

Blackol—unbleached kraft and black pigmented modified cashew nut oil resin.

Ceresine—a super-callendered unbleached kraft base impregnated and coated with an oleoresinous varnish.

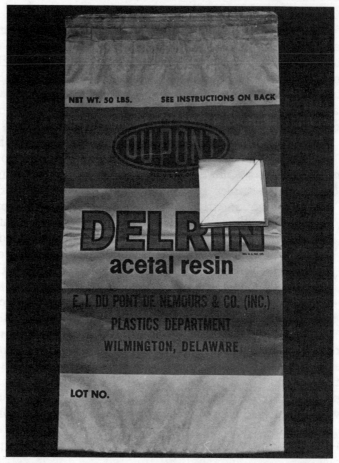

Courtesy of St. Regis Paper Co., Inc.

FIG. 2.7. LAMINATE INDUSTRIAL BAG

Pinch-bottom open-mouth with preapplied hot melt adhesive. Construction 1/20# Polyethylene-coated .00035 aluminum foil 6# Polyethylene 50 Natural Kraft; 4/50 Natural Kraft. Barrier properties of aluminum foil are (1) moisture, (2) odor, (3) vapor and gas.

Crystal cap—a bleached sulphite base coated with a white pigmented highly modified melamine, formaldehyde resin, rather a hard facing (not suitable for volatile solvents).

Flowed-in compounds—PVC or foamed materials, i.e., polyethylene.

Melinex film—50 or 100 gauge, rather hard (0.0005 or 0.001 in.).

Polyethylene—coated kraft, usually 0.002 in.

PTFE (0.002 in.) facing

Saran—0.00075 in. DOW PVDC film.

Solid Polyethylene

Solid PVC

Steran BK—coated PVDC.

Tinfoil Melinex—0.030 mm tinfoil (soft) + 50 gauge melinex.

In new materials (such as polyethylene, PVC, polypropylene) it is possible to combine the wad/facing as one material, utilizing the combined properties of resilience and resistance. Taking this one stage further, wadless caps can now be made from these materials. Caps can be fabricated from:

Metal—tinplate: corrosion protection can be improved by use of external enamels (which can be printed) and internal lacquers.

aluminum alloy—can also be enamelled, printed and lacquered.

Plastics—thermosetting: [i.e., phenol formaldehyde (PF); urea formaldehyde (UF)] These may be wood, paper or flour filled. Thermoplastic: Polystyrene (hard, rigid or brittle) can be embrittled by chemical attack; Polyethylene low-density, (L.D.), rather soft, flexible, can be wadless, may stress crack; Polyethylene, medium-density (M.D.) are high density (H.D.), less flexible and soft; Polypropylene (fairly hard) can be wadless if special sealing ring is incorporated and can crack at low temperatures.

The application of a prethreaded cap relies on a torque (now measured as kNm2 previous lb force in.) to make a good impression between two surfaces. Ideally one material should be hard (i.e., glass) so that it bites into a resilient material such as composition cork. The sealing surface of a glass finish may be radial or flat. In general (but not invariably) the unscrewing torque is 50 to 75% of the application torque. The torque instruments available are Torquemaster and Kork-a-torque. Unscrewing torques may however change according to storage conditions and length of time stored. Thermosetting caps change dimensionally (moisture and coeffcient of expansion) and metal/thermoplastic caps change mainly due to coefficient of expansion under changing conditions of storage. Phenol formaldehyde caps are slightly more dimensionally stable than urea formaldehyde caps. PF are available only in the darker colors, whereas UF is available in all colors including pastel shades. In addition to the more conventional screw caps in metal and plastic, other variables based on either the screw thread or lug principle are made.

Screw cap:

Double shell metal cap—gives a smooth external finish similar to a plastic cap.

Vacuum seal—(as used for home bottling of fruit, Kilner closure). Loose disc in metal or glass and sealing ring plus an external screw ring to retain the disc.

Courtesy of The Metal Box Co., Ltd.

FIG. 2.8. RANGE OF POLYPROPYLENE WADLESS SCREW CAPS
MADE TO BRITISH STANDARD R3 R4 AND METRIC MEDICAL
FINISHES

Lug cap:

Internal screw stopper—plastic plus rubber sealing ring, brewing and soft (carbonated) drinks.

Multi-start thread—Multi-start threads provide more even pressure on the wadding. Single-start threads may tilt or apply uneven pressure if the engagement is half a turn or less. One complete turn of thread engagement is ideal but a 3/4 turn or more is generally considered satisfactory.

British Standard 1918—lays down the standard for the R3/2 shallow, prethreaded closures and container finishes.

Unlike metal, plastics can be molded into a wide range of shapes and colors, thus adding to the decorative appeal of the pack. Thermosetting caps are compression molded either on platen or rotary-type equipment. Pressure, dwell time and temperature are critical to the curing operation, otherwise brittle caps will be produced. Thermoplastic caps are injection molded. Extra decorative effects can be added by blocking, embossing in the mold, inserts and metal vacuumizing. The closures used by such cosmetic manufacturers as Avon and Helena Rubinstein readily show what can be achieved in the way of decorative appeal. The cost of laying down injection molds is high, probably $600 to $1,000 per cavity. Compo cork is recommended for widemouthed closures as it takes up any unevenness (undulations) of the sealing surface.

Rolled-on and Rolled-on Pilfer-proof Closures

This process applies only to metal shells made from aluminum alloy. The caps may contain conventional wadding or a flowed-in gasket. In theory this type of closure provides a more consistent first seal than a prethreaded screw cap in that the operation of application: (1) Applies a top pressure which makes a seal between wad and container sealing surface. This top pressure can be controlled within a narrow torque range. (2) While still held by the top pressure, the thread is rolled into the metal, thus giving a finish which is tailor-made to the individual container. (3) In the case of a pilfer-proof closure the skirt is rolled under the bottle retaining bead. Disadvantages can occur with the R-O system where the glass finish is not up to standard. Stepped threads finning, etc. can cause perforation and difficulties in unscrewing. Reclosure is satisfactory provided the thread is fully formed (either on bottle or cap). Shallow formed threads can be made to override due to the softer alloy. Flavor-lok is a special form of roll-on closure used for minerals, carbonated drinks, etc.

Other Metal Closures for Glass and/or Plastics

The two foremost metal closures in this category are the crown closure (beer bottles) and the foil top as used for milk bottles. The crown cork closure is basically a shell which contains a wad/facing. This may be conventional, polythene or a flowed-in compound. Its application is similar to the R-O system in that top pressure is applied first, followed by application of crimping action. The metal used is tinplate. It is designed as a one-use closure. The foil milk bottle closure is unique in that it relies on the bottle being maintained in an upright position. Although it provides good protection against dirt and dust it is by no means an air or liquid proof closure (metal to glass seal without any wadding). The aluminum foil used is 0.04 mm to 0.065 mm.

Metal lever cap—Tinplate and wadding, can be resealed but is being rapidly replaced by Flavor-lok type closures.

Center pressure cap—Tinplate and wad. Press cap center for cap removal; press cap sides for retention. Used on glass jars but major use is on metal drums.

Metal vacuum seals—(1) Top sealing disc and rubber or plastic seal. This is held on by a safety band ring as used on jars of paste. (2) Light gauge aluminum cap incorporating a flowed in gasket. The cap is crimped under the neck ring, i.e., jam, pickles, etc. Omnia (continuous beading) and Garda (perforated beading). (3) More rigid closure usually made in tinplate applying a side seal on the neck by a rubber ring or flowed in compound.

Sterile seals—(1) Rubber plug with central skirt, plus crimped aluminum ring seal. (Goldy seals). (2) Combination type seal: rubber disc plus crimped ring seal. (3) Rubber flanged plug with central skirt, plus screw-on ring seal, as used for transfusion solutions.

Heat-sealed foils—The use of heat sealable foil has extended considerably in recent years. It is used for foil containers, blister packs, injection-molded containers, plastic, metal and glass bottles. With the last, Lectraseal is of special interest.

Other plastic containers—Plastic materials, particularly poly-ethylenes, have found special use as stoppers, plugs and snap-on type containers. This snap-on feature is also used on the Jay-Cap closure. The characteristics of plastic have also been used in various dispenser type of closures: for example, sifter and self-adhesive label (vim), spout and attached over cap, snip off spout with or without separate overcap, rotary closure for powders, rotary closure of tablets, pump appli-cators, flip tops.

Closures for Plastic Bottles—The more rigid plastics (PVC, H.D. Polyethylene, polystyrene, poly (4-methyl 1-pentene) (TPX), nylon, polycarbonate) will accept the same closures as glass. With the more flexible plastics the thread and caps may require modification of the thread shape to a buttress configuration (not suitable for metal caps). Multi-start threads provide an alternative means of capping. As with detergents, flip tops with an internal plug fitment are a popular form of closuring.

Other closuring materials

One of the earlier forms of closuring was a cork. This is still widely used for wines. Rubber corks and bungs are also used to a limited extent.

Summary

It has been indicated that closuring plays an important part with nearly every type of pack, although its role throughout may vary from one of retention with some degree of protection against dirt and dust to a hermetic seal (the perfect closure). The effectiveness of a pack is frequently associated with its closure, but, the barrier properties of the packaging material used may apply certain limitations. Poorly sealed packs can significantly reduce the shelf-life

Courtesy of Avon, Ltd.

FIG. 2.9. AVON PACKAGE
Cork is still used to convey a certain image in this gift package.

of a product. Incorrectly chosen barriers (including facing for wads) can adversely affect the product by such effects as contamination, corrosion, ingress of moisture, etc.

BIBLIOGRAPHY

ANON. 1973A. Statistics of glass packaging. Packaging *10*, No. 6, 17. (Swedish)

ANON. 1973B. Developments in packaging in glass. Pira seminar.

ANON. 1973C. Glass. National Center for Resource Recovery, Inc., Washington, D.C.

ANON. 1974A. Big-mouthed drinks bottle goes on test with rip-off cap. Packaging News, p. 1.

ANON. 1974B. The mechanization of the glass industry. Glass *51*, No. 1, 33–34.

ANON. 1974C. Kaaru—the glassware that took NZ by storm. N.Z. Packaging (Suppl.) No. 1, 9.

ANON. 1974D. Glass fibers from waste glass. Ind. Recovery No. 9, 11–13.

ANON. 1975. Energy price increases strengthen the position of glass. Glass View No. 3, 6–7.

ARRANDALE, R. S. 1974. How to pick a coating to help glass bottles resist shattering. Packaging Eng. No. 5, 54–56.

BRISTON, J. H. 1968. Progress in milk packaging depends on consumers. Inst. Packaging J. *14*, 24–26.

CHAMBERLAIN, P. 1974. Rationalization of bottles. Intern. Bottler Packer No. 3, 69–70.

CHILD, F. S. 1973. Four products using waste glass as raw materials found to be economically viable. Am. Glass Rev. *94*, No. 2, 12,20.

CIANFLONE, F. 1974. Glass as packaging. Tech. Imballagio No. 5, 149–154. (Italian)

COOK, R. 1973. When a bottle's not a bottle. UG Packaging, pp. 8–9.

CORNAZ, M., and JONES, S. P. 1975. Present state and future possibilities for process control in the glass container industry. Glasstech No. 3, 51–55. (German)

DOREY, R. 1973. Upgrading the strength and performance of glass containers. Glass *50*, No. 12, 358–360.

FOSTER, T. V. 1973. The future of forming. Glass Technol. No. 6, 157–162.

GOOCH, J. U. 1974. Packaging in Glass. Pira Bibliography, 29 pp.

GOODING, K. 1974. How short is a shortage? Glass View *1*, No. 1, 3–4.

JONES, N. 1974. Glass fights back. Australian Lithography *8*, No. 47, 49–50.

KREIDEL, N. J. 1974. The state of glass technology, assessment and outlook. 10th Intern. Congr. Glass No. 4, 82–92.

PARDOS, F. 1974. International comparisons of packaging industries. Emballages No. 319, 122–133, 135–136, 139–141, 143–145, 147–151. (French)

SCHARFENSTEIN, O. H. C. 1973. Glass containers successful in the fruit and vegetable conserve sectors. Neue Verpack. *26*, No. 9, 1348–1349.

SVEC, J. J. 1975. Glass is the lowest cost container. Ceram Ind. No. 2, 15–16.

WEST, F. G. 1975. The release of toxic metals from glasses and glazes. Glass No. 1, 19–20, 22–24.

WIDOFF, E. 1973. Uncertain whether glass will be clearly that glass. Packaging *10*, No. 6, 15–32. (Swedish)

CHAPTER 3

Metals

INTRODUCTION

Metals have been known for thousands of years by man. Myths have been built around metals and wars have been fought for their control. An ancient Peruvian tradition claims that the orders of society came from eggs of different metals—golden for the chiefs, silver for the nobles and copper for the common people.

Metals Known in Prehistoric Times

Six metals were known and used by prehistoric man: silver, gold, copper, tin, lead and iron. Most were found free in nature, occuring naturally in pure metallic form. Silver and gold, too soft to be of value as tools, were prized as ornaments probably because of their beauty. Copper and tin were mixed together and shaped into tools such as hammers and spear heads. These bronze tools replaced stone and flint because the bronze was so much more durable. So important was bronze to the ancients that historians have named the time of its greatest use the Bronze Age. Lead, more easily worked than copper and more abundant than silver, was shaped into vessels and later rolled into pipe form. Meteoric iron, when it could be found, was the most valuable of all, particularly since it contained small quantities of another metal, nickel. Iron itself is the basis on which the steel industry is based. Primarily an alloy of iron and carbon, steel is the most common metal in use today. It can be produced with a wide variety of useful characteristics, most of which are related to strength and toughness. Pure metallic iron, however, is rather soft and fairly easy to work, but it is in this form that it was known to early man.

Metals Discovered During Early History

At least eight other metals were discovered and used during the eighteen hundred years of the Christian era. These included mercury, one of the only two metals which are liquid at room temperature. (The other is gallium, a rare metal with limited use in high-temperature thermometers). Mercury was in use by the 8th century and was a favorite material for experimentation by the alchemists.

41

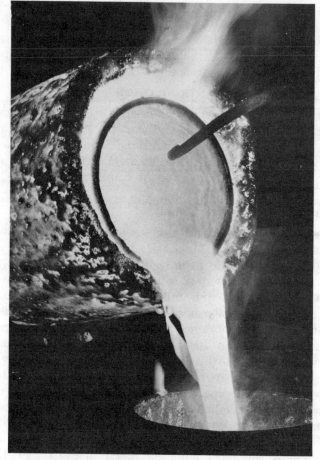

Courtesy of South African Information Service
FIG. 3.1. PRECIOUS METAL—GOLD BEING POURED

Zinc was recognized as a metal during the 16th century. Copper and zinc form another alloy, brass, a generic name for an entire family of alloys, each member of which contains copper and zinc in different proportions. One of the more important uses of zinc today is for galvanizing, in which sheet iron is coated with a thin layer of metallic zinc, primarily to prevent the iron from rusting.

Antimony, bismuth and arsenic were also recognized as metals during the Middle Ages. They are closely related chemically and are similar in physical properties. Antimony is the principal alloying ingredient added to lead in making type metal. This metal is different from most other metals in that it expands slightly as it cools after being poured into a mold, and so fills even fine detail in the mold.

The alchemists who worked with antimony probably confused it with lead. Probably confused with tin was bismuth. One of the primary uses of bismuth is in the production of low-melting alloys, some of which liquify well below the boiling point of water. (The fusible plug in automatic fire-sprinkler heads is usually a bismuth alloy.)

Arsenic has been known for centuries primarily as a poisoning agent, in earlier times for people, but more recently for insects. Arsenic as a metal has limited use as a hardening agent in the production of lead shot.

Nickel has been in use for thousands of years, but has been known as a separate metal only since 1750. Alloyed with copper in the manufacture of coins and medallions in ancient Greece, it is still an ingredient of coins in many countries today. Nickel added to basic steel imparts toughness and strength. Stainless steels contain up to 35% nickel.

Manganese is a hard, brittle metal with a reddish tinge to its silvery luster. It was recognized as a separate element in 1774. Almost nine-tenths of the manganese produced today goes into alloying of steel, to which it imparts toughness and hardness. Steel for such applications as plow blades may contain up to 12% manganese.

Another of the so-called precious metals, platinum, occurs free in nature, but was not known in Europe until 1735. It quickly became of value beyond its use in jewelry because of its resistance to corrosion by strong acids, whereby it could safely contain experimental mixtures in the chemical laboratory. It has found increasing use in recent years as a catalyst.

Recently Discovered Metals

The more recently discovered metals, identified or isolated since about 1800, generally have names ending in "um" or "ium." Many of these are of commercial importance because of their use as alloying ingredients with more common metals.

Beryllium was not discovered until 1797, and even now is fairly difficult to find. Its main source is the mineral beryl, which is produced only as a by-product in mining other minerals. Beryllium is added to copper and to the brasses in the production of springs for the electrical industry. In the development of nuclear energy beryllium serves as a moderator for the generation of slow neutrons.

Cadmium is a silvery metal with a bluish tinge when pure. Discovered in 1817 it occurs in small quantities in zinc ores and is generally produced as a by-product of zinc. Cadmium is alloyed with bismuth to produce some of the low-melting metals referred to above. As the pure metal, cadmium is used as a protective coating for nails, bolts and nuts. It is added to silver to increase the hardness of dental fillings and to prolong the life of silver plate.

Four of the more recently discovered metals, chromium, tungsten, vanadium and molybdenum, serve mainly as alloying additions in steel production. Of these chromium is probably the most evident since it helps produce the high luster of stainless steel. Up to 30% chromium, usually in combination with nickel, goes into this rust-resistant alloy. In much smaller quantities chromium increases the hardness and strength of steel.

Tungsten has been recognized as a metal since 1783, but little use was found for it until the development of the incandescent lamp. Tungsten proved to be an ideal filament material and is still so used. As an alloying metal in the steel industry, its extremely high melting point gives tungsten-steel cutting tools the ability to operate even at red heat. It also adds strength and hardness to armor plate.

Vanadium, discovered in 1830 in certain ores, is added to steel to increase fatigue resistance and toughness. It is the hardest of all the metals. Vanadium is difficult to purify, but comparatively easy to produce in the form of alloys with iron and nickel. Oddly enough it is more abundant in the earth's crust than either zinc, nickel or copper. Vanadium steel is primarily used in the production of drills and other woodworking tools.

In contrast to the abundance of vanadium, molybdenum is one of the scarcer metals in the earth's outer skin. Like tungsten it adds heat resistance to steel cutting tools. In close association with tungsten it serves as a support for filaments in incandescent lamps. Molybdenum is frequently added to chromium and nickel as an alloying metal in stainless steels. The pure metal was first separated from its ore in 1782. Gun barrels, armor plate and armor-piercing shells are usually made of steel containing up to 0.4% molybdenum.

Most of the metals occur on earth in the form of minerals buried deep underground. At least one, however, in actual commercial practice is harvested from the sea. Magnesium is present in sea water to the extent of 1300 ppm. Magnesium is one of the lightest metals, being approximately one-third lighter than aluminum. Like so many metals it is relatively useless in pure form, so it is alloyed with aluminum, manganese or zinc for many purposes. Perhaps the most important use of magnesium alloys is in the aircraft industry, where the saving in weight can justify the additional cost of the metal.

Although titanium was discovered about 1790, only within the last few years has it become important as a structural metal. This is partly because of the difficulty of freeing it from oxygen and

Courtesy of The Anaconda Co.

FIG. 3.2. REFINERY—CASTING COPPER WIRE BARS

nitrogen, and partly because it is so difficult to work. When freed from impurities titanium becomes a hard strong metal, capable of resisting heat at least to 400°C, whereas aluminum and magnesium lose their strength. It would be in much wider commercial use were it not for its high processing costs. Titanium is added in small quantities (up to about 0.2%) to some aluminum alloys to increase heat resistance.

In addition to platinum and gold, the so-called precious metals include iridium, rhodium and palladium. Physically they are similar to platinum, being heavy, hard and chemically inert. They are all very expensive, ranging in value from about $25 per oz for palladium to nearly $200 per oz for iridium. They are used in such items as ball-point pen tips, electrical contactors in telephone dialing equipment, and pivot points for chemical balances. Platinum-iridium alloys, because of their resistance to food acids, are valuable in dentistry.

Aluminum is a member of a family of metals that also includes gallium, indium and thallium. Like magnesium, pure aluminum is too soft to be of structural value. Alloying with silicon, magnesium, manganese, copper and zinc produces metals with a wide range of physical and electrical characteristics. Aluminum is roughly one-third as heavy as other common metals (iron, zinc and copper) and about 50% heavier than magnesium. It is chemically quite active and would corrode rapidly in moist air were it not for a film of oxide which forms almost instantly on a freshly-cut aluminum surface. Aluminum oxide prevents the further corrosion of the metal beneath it.

Since the modern era of packaging dates from about the time of the industrial revolution of the 17th and 18th centuries the metals we need concern ourselves with today are almost exclusively steel, tinplated steel, chromium coated steel and aluminum. Gold, silver, platinum, pewter, brass and copper are now too expensive for any but ornamental and more permanent uses. Tinplated sheet iron was developed in Bohemia in 1200 A.D. and kept a jealously guarded secret for several hundred years.

Aluminum was not isolated until 1825 and remained a rare curiosity until the late 1880's when a practical method of extraction from bauxite was developed. By the late 1920's and early 1930's it began to compete with copper and steel for household items such as cookware. Its widespread adoption in packaging came in the mid-1940's when techniques for rolling and decorating thin aluminum foils were perfected.

TYPES OF METALS AND THEIR MANUFACTURE

Steel

Steel is one of the two major substrates used on rigid metal packaging. Prior to 1850 steel was made by a method known as the "crucible process." In this process a charge containing iron was placed in a refractory crucible which was placed over a fire. If the fire was hot enough the charge eventually melted. It was then stirred constantly to expose more of the surface to the air so that the impurities would oxidize and float to the surface. This process is confined almost entirely to the lower melting point metals such as lead, tin and copper.

The crucible was replaced in about 1860 by the Bessemer converter, the invention which perhaps was the greatest single

Courtesy of Reynolds Metals Co.

FIG. 3.3. BAUXITE MINING

influence in creating the Steel Age. The Bessemer converter operates on a simple principle. Molten pig iron plus scrap steel and iron ore are mixed in a large pear-shaped vessel open at the top and which has means for introducing a large quantity of air under pressure from the bottom. Despite the fact that this air is cold, oxidation reactions within the converter produce enough heat to raise the temperature of the melt to the point where both silicon and carbon separate from the charge. The silicon forms a slag which is later removed, while the carbon is burned off as carbon monoxide, resulting in the 25-ft flame typical of the Bessemer process. Pig iron containing about 4% carbon is converted to steel whose carbon content is about 0.06%, while the silicon content is reduced from 1.2 to 0.03%. At the same time manganese present in the pig iron in quantities up to 1% has been lowered to 0.05% in the steel. All these reactions result from blowing air through the melt with no external heat added except in the molten pig iron. This relative simplicity helps to make the Bessemer conversion an economical operation. With modifications it is used to produce millions of tons of steel each year in the United States.

The success of the Bessemer process for steel-making and the increased demand for steel generated research into alternate methods of production. In the middle 18th century Siemens invented the open-hearth process or the reverberatory furnace. Almost any nonferrous metal that can be smelted in a blast furnace can also be handled in the reverberatory furnace, although relatively high heat losses restrict the maximum temperature that it can reach. Far outweighing this, however, is its ability to handle a charge of almost any physical characteristics. Very coarse or very fine ores can be processed with little difficulty and the chemical composition of the products can be closely controlled. The charge is relatively shallow and spread out over a wide area. Flame from the burners plays over most of the surface of the charge, while heat reverberates, or bounces back, from the roof of the furnace. No fuel is included in the charge as in the blast furnace.

The reverberatory furnace is currently the most important of the refining furnaces, processing as much as 85% of the world's steel. Several reasons exist for the widespread use of the open hearth furnace. The furnace itself is not, for practical purposes, limited in size as in the Bessemer converter. The use of the so-called regenerative system permits higher temperatures and appreciable savings in fuel. The hot flue gases are forced through a checker-work arrangement of refractory tile built at either end. During part of the heating cycle the hot gases flow out one end of the furnace giving up much of their heat to the tile. Later the flow is reversed and the incoming air and fuel is heated by the tile. In this manner little operating heat is lost. Since the reactions proceed at a much slower rate they may be more closely controlled. The few disadvantages of the reverberatory furnace, such as higher initial cost and lack of agitation of the melt, are more than offset by its many advantages. As a result it is the most widely used, not only for steel refining, but for many other metals as well. Almost all of the commercial tonnage metals are, or can be, treated in such a furnace.

Steel is also made by the electric arc furnace (Heroult-type furnace), the Basic Oxygen (Linz-Donawitz) process and the continuous casting process, all of which yield a fine quality steel.

Tinplate Steel.—Tinplate steel is low carbon steel coated with a thin protective layer of tin. The use of a tin coating generally protects the base steel from rust and corrosion. Two methods are used to produce tinplated steel: hot dipping and electrolytic. Prior to World War II practically all tinplate was made by drawing sheets of steel through a bath of molten tin to produce a hot dipped tinplate. Now more than 95% of all tinplate is made by electroplating tin on

Courtesy of United States Steel Corp.

FIG. 3.4. ELECTROLYTIC TINNING LINE

Tin is applied to steel strip, previously rolled to a thin gauge on this electrolytic tinning line.
It may be used for food or beverage containers, toys or kitchen utensils.

steel at the steel mill. A coil of steel plate is unwound through an electrolyte. Pure tin bars acting as anodes are inserted in the electrolyte. When an electric current is passed between the anode (tin bars) and cathode (coil of steel plate) deposition of tin takes place. From Faraday's Laws of Electrolysis the amount of deposition can be controlled by varying the current, the speed of the sheet through the cell or the condition of the electrolyte. Depending on the type of electrolytic line used, this plating may take place at speeds up to 1200 ft per min. It is also possible to vary the tin coating on one side of the plate while holding the other constant to produce a differentially coated plate. The manufacture of tinplate is a highly complex operation involving extensive initial capital investments.

Other steel-based materials which may be used in packaging include steel itself, galvanized steel, terneplate and blackplate. Galvanized steel is steel which has been hot-dipped in an alloy of 80% lead and 10% tin. Due to the lead content terneplate cannot be used for food applications. Blackplate is simply uncoated tinplate base. The name blackplate comes from the fact that some of the early production was covered with black iron oxide. Tinplate prices have been pressured upward by the soaring cost of Malayasian tin.

Tin's price fluctuations have provoked interest in emerging use of tin-free steel and blackplate for two-piece drawn-and-ironed cans in the beer and soft drink market.

Tin-free Steel (TFS).—During the 1960's tin-free steel (TFS) was commercially developed for use by the canmaking industry. The base steel used is the same as for tinplate. Instead of tin, however, an alternate coating is applied on the steel base. The coating may be chromium-chromium oxide, aluminum or a phosphate-chromium film (chemically passivated steel). Chromium-chromium oxide is coated on a mild steel base in the range of 0.2 in. thick, almost 1/30 the guage of an average tinplate coating. While this coating will not protect the steel from strong acid or alkali attack, it serves as a satisfactory barrier to weak acids and alkalis. Also available is the earliest TFS, steel made with a phosphate-chromate type of passive film on its surface. The stock has only limited application and is being used to fabricate large containers.

Protective Coatings

Cans were originally plain and had no interior protective coatings. It became apparent that many more products could be packaged if a coating were put on the inside of the can. Such a coating could minimize the chemical action between the product and the tin-coated steel. One example is the reaction between acid fruits and tin:

$$\text{Acid Product (pH} < 7.0) + \text{Tin} \rightarrow \text{Tin Salt} + H^2 \uparrow$$

This reaction liberates hydrogen which would create a pressure inside the can and bulge it. The reaction would also impart a metallic flavor to the product due to the tin salt present.

The effort to neutralize these internal reactions and to minimize exterior corrosion has led the can industry to adopt more than 50 different protective coatings. Four main types of enamels are used in the container industry today. These are named for the type of solid resin used to make up the enamel: oleoresinous, phenolic, vinyl and epoxy. Besides the basic resin blends other constituents are added to make the finished enamel, for example, drying oil to aid polymerization (solidifying process), and dryer to accelerate this polymerization with solvent as a resin vehicle. In addition the exterior surface of the container may be coated a color design such as in beer cans or with a coating of clear varnish to protect the exterior from atmospheric corrosion.

Aluminum

Aluminum sheet is used in the fabrication of both multipiece

Courtesy of The Metal Box Co., Ltd.

FIG. 3.5. TINPLATE FOOD CANS ENTERING THE TESTER

Cans are subjected to test by air pressure to establish whether
there are any potential leakers.

containers and formed one-piece container bodies. Aluminum is the world's most abundant metallic element comprising about one-twelfth of the earth's crust. Aluminum exists, however, only in chemical combinations with oxygen and other elements. The production of foil like that of all other aluminum products starts with the mining of bauxite, an ore containing 40 to 60% aluminum oxide.

TABLE 3.1
TYPICAL COMPOSITIONS OF ALUMINUM

Clay:	Kaolinite $Al_2O_3 \cdot 2SiO_2 \cdot 2H_2O$	+	Impurities FE_2O_3, TiO_2
Bauxite:	Gibbsite $AL_2O_3 \cdot 3H_2O$	+	Impurities Fe_2O_3, TiO_2, SiO_2

FIG. 3.6. ALUMINUM PRODUCTION

Molten aluminum is tapped from first of a series of electrolytic smelters which began operating July 9, 1959 at Reynolds Metals Company's aluminum reduction plant at Massena, N.Y. The two bays of smelters seen here are part of the first potline. The three lines produce 100,000 tons of aluminum per year. Metal from this first tap will be delivered in molten form to nearby Chevrolet aluminum foundry.

Separating the aluminum oxide from the other elements in bauxite involves a long, complicated, chemical process. The resulting product, alumina, is a white powdery material that looks much like coarse granulated sugar. Alumina is dissolved in molten cryolite at 1800° F and then separated into oxygen and metallic aluminum by the passage of electric current. Molten metal accumulated at the bottom of the reduction pots is siphoned off into ladles at regular intervals, the frequency depending upon the size of the pot.

About 4 lb of bauxite are required to make 2 lb of alumina, which when broken down result in 1 lb of metallic aluminum. The reduction operation requires almost 8 kwh of electricity per pound of aluminum. After reduction the molten aluminum is poured into rectangular molds to cool and solidfy. These chunks of raw material, called ingots, may weigh from 30 lb to a maximum of 15,000 to

20,000 lb. Before being cast into rolling ingots the metal is purified and in some instances alloyed. Several operations are performed on the ingot prior to rolling including scalping, homogenizing, and reheating, usually in that order. Most freshly cast ingots contain many undesirable surface blemishes, such as small tears and oxide inclusions. In large milling machines an operation known as "scalping" completely cleans both faces of the ingot, makes then flat and removes surface porosity so that only sound, clean metal remains.

Homogenizing is a thermal treatment which relieves the internal stresses resulting from casting and uniformly distributes the various alloying constituents throughout the ingot. Temperatures ranging from 950° to 1050°F for a period of about 24 hr are employed for homogenizing. Some alloys are so pure that this process is unnecessary, making it possible to bypass the homogenizing step.

Before rolling, the ingot must be reheated to raise its temperature to about 850° or 900°F, at which point it becomes plastic and can be rolled easily. An average of 8 to 12 hr reheating time is usually required to assure uniform temperature throughout the entire ingot. The time depends upon the alloy and size of the ingot. After reheating the ingot is taken to the first of several hot-rolling mills for further processing.

Current practice in many plants is to begin "working" operations to produce sheet (a solid section rolled to a thickness range of .006 in. through .249 in.) by passing the ingot between two vertical rolls which "edge roll" it. This process squares up the two long edges and starts the "kneading" action which breaks up the as-cast structure and eventually produces the "wrought" structure desired as rolling is continued. Mills used for rolling sheet are designated by the number of vertical mill rolls in their construction. They are called 2-Hi, 3-Hi and 4-Hi. Four-Hi reversing mills are generally used for the initial or "breakdown" rolling of the heated ingot after it has been edge rolled (although 2-Hi mills are sometime used). The ingot is passed back

TABLE 3.2
PRIMARY ALUMINUM PRODUCTION—
MAJOR RAW MATERIAL REQUIREMENTS

Material	Per lb Aluminum
Alumina	1.93 lb
Petroleum coke	0.4 lb
Coal tar pitch	0.12 lb
Bath materials (fluorides)	0.05 lb
Power	8 kwh

and forth between the two small hard work rolls. An important characteristic of rolling is that the smaller the diameter of the work rolls, the larger the degree of reduction since the pressure per unit of area transmitted from a small roll is considerably greater than that from a large roll. Reversing mills currently used by the aluminum industry are as wide as 160 in.

Next the width of the sheet is controlled by "cross rolling" (rolling the slab crosswise) to widen it. Thus, the first series of passes are made by cross rolling the slab until the desired width is attained, adding a few extra inches for trim allowance. Then the slab is turned 90 degrees and rolled lengthwise for second series of passes. As material being rolled elongates only in the direction of rolling, the increase in width during lengthwise rolling is negligible. The amount of thickness reduced (or reduction per pass) will vary according to the alloy, width, and thickness of the material being rolled. A typical reduction on relatively soft material (such as alloy 3003) is 1-1/2 in. per pass. Water soluble lubricants are continually sprayed on the work during all hot-rolling operations. The breakdown lengthwise rolling passes are generally continued until the slab is only a few inches thick and 20 ft or more in length. The irregular ends are then sheared off and the edges trimmed as required. These breakdown operations greatly elongate the ingot and considerably reduce its thickness. Rolling under very high pressure causes the metal to work-harden and it is necessary to remove internal stresses by reheating at this stage. The slabs are then further reduced by additional hot rolling to plates about 3/4-in. thick.

After being thus reduced in thickness the material is subjected to continuous hot rolling in a series of 4-Hi tandem mills. Tandem means a series of mill stands erected close together and operated as a single unit. Tandem mills differ in principle from breakdown mills only in that they operate continuously in one direction; that is, a sheet passes through several mill stands at the same time, each succeeding and stand going faster than the previous one to accomodate the elongating metal. As the speed of the stands increases progressively from the entrance to the exit mill, the sheet moves through them very rapidly and must be kept under high tension. Upon emerging the sheet has usually been reduced to a thickness of about 1/8-in., completing the hot-rolling sequence.

Most of the material is coiled immediately after hot rolling, although it is sometimes allowed to cool first on a long run-out table. At this stage the coils are annealed to relieve built-up stress before cold rolling.

Cold rolling is performed on equipment very similar to that used in hot rolling. The cold-rolling process must be controlled more carefully than hot rolling because of the closer tolerances required. Cold rolling requires greater power because the cold metal is harder to work. In addition the rolls of cold mill stands are ground and polished more finely than those of the hot mills. A mineral oil is utilized as a coolant and lubricant in cold rolling instead of a water-soluble type oil used in hot rolling. Oil is a critical factor in cold rolling and must be kept clean and carefully controlled for quality production. After the coils from the hot mills have cooled for about 36 hr they are cold rolled in single or tandem 4-Hi mills. The thickness of the material is reduced about 50% per pass. Thus a 4-Hi two-stand tandem mill can reduce thickness by 75% in one pass through the two stands. Single-stand mills are used primarily for lighter thicknesses and usually for the narrower widths. Cold-rolling operations must be carefully planned as the quantity of metal rolled to a certain thickness may require annealing before further rolling can be accomplished. Great flexibility must be provided in scheduling both materials and equipment, as different alloys require different rolling sequences.

Sheet to be used in the rolling for foil is generally cold-rolled to 0.125 in. thickness. It is then wound in coils ranging from 5,000 to 15,000 lb. Coiled sheet (commonly known as "reroll stock") is in the "as-rolled" or approximately full hard temper and has an oily surface to protect it in transportation. Sheet to be used in container manufacture is rolled to the desired initial gauge. In the case of formed semirigid foil containers, the final gauge of the finished container ranges 2.5 to 8 mils depending on size. In such containers there is very little change in gauge in the forming operation. In rigid formed containers such as drawn cans, however, body stock sheet gauges run from 9 to 10 mils and can end sheet stocks about 13 mils. In the drawing process the gauge is substantially changed.

Properties of Various Types of Metals

Steel, Tinplate and Tin-free Steel (TFS).—Steel is generally selected for canmaking by its temper (i.e., degree of hardness). As the steel temper number increases from T1 to T6, temper correspondingly increases. A deep drawing application would require a low temper T1 steel while a shallower draw would utilize a T3 temper. Types of steel currently available are: (1) MR—container steel; (2) L and LT—used for the production of tinplate and also resistant to corrosive foods; (3) MC—rephosphorized steel for can ends for non-corrosive products; and (4) D—deep-drawn steel for deep drawing application. Tinplate is produced from these tempers and used in can fabrication. Lightweight double-reduced 2CR tinplate (double cold-reduced plate) is used primarily for beer and carbonated beverages. Thinner than normal tinplate, 2CR plate is strengthened by a process which provides great stiffness. Three tempers are available in 2CR: DR-8, DR-9 and DR-10 (hardness increases with the last digit).

The unit of gauge in canmaking terms is LBS/BASE BOX or (pounds per base box). A base box is designated as 112 sheets of any uniform gauge, each sheet having dimensions of 14 in. × 20 in. The weight of this base box, there, defines the gauge; weight per base box and usually varies between 75 and 118 lb. Since LBS/BASE BOX is proportional to gauge the conversion is approximately:

Gauge (in.) = Plate Weight (LBS/BASE BOX) × 0.00011 in.

Another important consideration in the basic steel is its chemical composition, which influences the hardness and corrosivity of the steel. The canmaking industry specifies plate weight (thickness), chemical composition tolerances, temper (hardness) and tin coating weight for all metal cans packed with products. These specifications are derived from through actual storage tests of prodcts to be packed in various container specifications. The amount of tin deposited on the steel plate is also defined as LBS/BASE BOX and usually varies between 0.25 to 1.25 LBS/BASE BOX total on both sides of the plate.

Aluminum.—*Alloy.*—Aluminum is normally supplied in the high purity grades (1000 alloy series) but other alloys are now being used for special types of products.

Most widely used in the packaging industry, 1235 alloy has a minimum content of 99.35% aluminum. It is commonly referred to as foil analysis. Unless there is a specific alloy requirement in the end use, 1235 alloy should suffice. Having a 1.0 to 1.5% manganese content, 3003 alloy is used in the fabrication of formed or drawn containers and also in forming various caps and closures.

Temper.—Aluminum is produced in "O" and "H" tempers. O temper is produced by subjecting the material (after it has been rolled to the desired thickness) to controlled heat followed by controlled cooling. This treatment relieves stresses created during rolling and allows crystal realignment. Aluminum with O temper, commonly referred to as dead soft, is the softest form available and has the lowest physical strength. It has good folding and forming characteristics, however, and therefore is widely used in converting operations.

Temper is produced by work hardening the metal such as in the normal foil-rolling operations. H12 and H14 tempers are strain hardened by rolling to temper after an intermediate anneal. This improves the physical characteristics over dead soft, but controlling the physical properties of intermediate tempers is difficult. They are not recommended where definite properties to close tolerances are involved. H18 is strain hardened by rolling almost to the maximum limits. Foil rolled without intermediate or final annealing is in approximately this temper and is considered the direct opposite of dead soft foil. H19 is fully strain hardened by rolling to give maximum physical properties. Some fabricating end uses require physical properties greater than ordinary so this supperhard aluminum foil was developed. H24 is an intermediate temper, strain hardened by rolling, and then partially stress relieved by annealing. Only a limited percentage of the total production of foil is presently used in hard or intermediate tempers, most being in the soft or O temper.

In the rolling and annealing operations, various surface conditions may be produced: (1) Dry annealed—The surface is free from an oily film. The foil is annealed in such a way as to dissipate the residual rolling oil. The dry surface is suitable for lacquering, printing, or coating. It can be wetted by water-dispersed adhesives. (2) Slick annealed—The surface has a slight film of oil. The foil is annealed in such a way as to prevent the complete dissipation of the rolling oil. The slick surface is used where better corrosion resistance than that provided by dry annealing is required. (3) Washed—The surface is substantially free from an oily film. During the last rolling pass, the foil is solvent-flushed by a washing agent which is used instead of the usual rolling oil. (4) Chemically cleaned—The surface is free from all rolling oil. After the last rolling operation the foil is passed through a chemical bath. The completely dry surface is suitable for critical converting operations requiring a cleaner metal than that produced by washing, for example, when a coating is to be applied. (5) Stearic acid coated—The surface has a wax-like film which is produced by passing the foil through a bath of stearic acid in solvent solution after the final reduction pass. Foil coated with stearic acid is used in fabricating operations. (6) CFA (coconut fatty acid) coated—The surface has a slight oily film produced by passing the foil through a bath of coconut fatty acid after the last rolling pass. CFA coated foil is used for difficult forming operations where maximum lubricity is required. (7) Wax textured—The surface has a film of wax, which is produced by passing the foil through a bath of wax (usually paraffin) and then chilling the wax. The operation is performed after the last rolling pass. Wax-textured foil is used for light machine applications such as bottle caps.

Courtesy of Reynolds Metals Co.
FIG. 3.7. ROLLING MILL
World's largest aluminum hot rolling mill rolls aluminum sheet at Reynolds
Alloy plant.

Aluminum has advantages over tin, lead, and steel because it is easier to roll, can be rolled thinner, and provides more footage per pound. Aluminum has a number of properties which account for its use as a material for can manufacture.

Reflectivity: The reflectivity of a pure aluminum surface ranges from 90 to 95% for infrared radiation and 80 to 85% for white light. These reflectivity values are not appreciably affected by roughening of the foil surface, but coatings will lower them by absorbing some of the radiation. The heavier the coating and generally the darker the color (particularly pigmented colors), the lower the percentage of reflectivity and conversely the higher the percentage of emissivity. This high reflectivity affords superb insulative and eye-appeal values that are not approached by any other packaging material.

Emissivity: The low emissivity of aluminum is utilized in foil cook-in or heat-in packages for prepared foods for which it is desirable that a minimum of heat be lost before serving. Emissivity is expressed as the ratio between the quantity of heat radiated from the surface of a given body and the quantity that would be radiated by a theoretical black body of the same dimension and temperature. Actually this will be 100 minus the reflectivity in percent. The theoretical black body is considered to absorb all radiant energy impinging upon it and then lose it by re-radiation back into its surroundings.

Thermal conductivity: Aluminum has higher heat conductivity than all other packaging materials. Of all metals it is exceeded only by gold, silver, and copper in this property. Aluminum permits faster freezing or faster heating by contact. Conductivity is defined as the transfer of heat from one part of a body to another part of the same body, or from one body to another in physical contact with it. It is generally expressed in terms of the quantity of heat flowing through a material in a given period of time under specific conditions. The dynamics of heat transfer are always involved with the flow of heat from the warm to the cold body or bodies.

Light weight: Aluminum is noted for its lightness in comparison with other commonly used metals. Steel weighs a little less than three times as much as aluminum. The use of light weight aluminum containers reduces shipping costs.

Corrosion resistance: The corrosion resistance of aluminum is excellent when compared with low carbon steel, and in some instances is greater than that of stainless steel. Copper resists some chemical attacks to which aluminum is susceptible, but in many instances aluminum is superior. A similar situation exists with other competitive metals in varying degrees. Some foods, especially those of high moisture and condiment content, react with aluminum even at room temperature. In such instances a protective coating is used to prevent contact.

Workability: Aluminum has excellent workability and is readily formed by all the various cold and hot methods applicable to other metals. It can be extruded, rolled to extremely thin sheets, and may be joined by brazing and welding as well as mechanical means.

Nonsparking: Aluminum does not throw off sparks when struck with other materials, whereas ferrous metals do. The advantages of this property of aluminum are self-evident where inflammables or explosive atmospheres are involved, such as may be found in operating rooms.

Moisture-vapor resistance: In bulk aluminum is impermeable to moisture vapor. In packaging a low water vapor transmission (WVT) rate is often a paramount requirement. (WVT is sometimes expressed as "moisture transmission.") Pinholes are extremely minute (for example, in foil .0004-in. gauge pinholes may range in size from .00000001 to .00003 sq in.). Such holes are usually filled in or blocked when the foil is laminated.

Gas resistance: The resistance of aluminum to gas is the same as its resistance to water vapor. By the proper selection of laminants the gas permeability through pinholes in the foil can be reduced to negligible amounts.

Grease and oil resistance: Aluminum is a complete barrier to grease and oil and will not absorb these substances.

No taste and odor: Being completely tasteless and odorless, aluminum can be used with the most sensitive materials. Aluminum packaged medicines or cosmetics will not take on tastes and odors of items that may be stored in close proximity.

Heat and flame resistance: Aluminum will not burn and is not affected by temperatures normally used in packaging applications.

Opacity: Completely blocking light rays, aluminum in combination with sound food preservation techniques will prevent rancidity and other light-catalyzed losses of flavor, vitamins, and color in foods.

Nontoxicity: Aluminum is nontoxic and therefore may be used in direct contact with food.

Strength: Although pure aluminum is very soft and ductile, it is capable of attaining high tensile strength by alloying, cold working and heat treating. (In foil packaging, however, desired tensile strength is obtained not so much by selecting a certain alloy and temper as by controlling the thickness and combining it with other materials such as paper and film.)

Tube Metals.—The collapsible metal tube was among the very first containers in wide use to be made of metal. John Goffe Rand extruded the first metal tube of tin in 1841 for the specific purpose of containing artists' oil colors. He produced a convenience package that, even with many refinements over the years, structurally is the same now as it was 115 yr ago. For many years virtually the only packaging application of metal tubes was for colorant. While the so-called compressible tube was the invention of an American, it was not until 1870 that the first tube manufacturing plant was established in the United States. The impetus for the tube industry's great growth after the turn of the century was the packaging in the 1890's of toothpaste in metal tubes instead of glass jars. The amazing success of tubed toothpaste prompted other manufacturers of paste or semi-fluid products to turn to the metal tube.

One of the great growths in collapsible tube packaging has been in the pharmaceutical field. The discovery and development of antibiotics and the acceleration in research in this area since World War II brought the collapsible metal tube into new prominence. It made possible the containment of ointments and other preparations made from the so-called wonder drugs without loss of potency or contamination.

The collapsible tube industry looks forward to continued growth, a growth that has remained unchecked despite stiff competition from other types of containers. One factor is consumer convenience, and another is its adaptability to virtually all forms of modern merchandising. There are many features that make collapsible metal tubes an ideal package: they are nonporous, light in weight, sanitary, durable, versatile, nonrefillable, decorative, easy to handle, allow a long shelf-life, and are adaptable to mass production methods and to automatic packaging. Tubes are eminently suited for dispensing in whatever quantities desired medicinals and pharmaceuticals, cosmetics, shaving creams, dentifrices, spread-type food products, and household and industrial items such as lubricants, adhesives, compounds of various kinds, colorants, dyes and a variety of similar products. Special applicators enable tubes to be used for safe and convenient application of medicaments to the eye, nasal passages and other body orifices. Similar utility is found in tubes for veterinary application, especially in the treatment of mastitis. A number of special tips and applicators are available for applying adhesives, lubricants, inks, caulking compounds, putty, etc.

Collapsible Tube Metals.—Metals used in the manufacture of callapsible tubes include aluminum, lead and tin. These materials withstand extreme cold working, a requirement eliminating most other metals. Cold working strengthens and hardens the metals, particularly aluminum, which is finally annealed to impart the necessary pliability. The properties of aluminum that make it collapsible tube metal are its lightness in weight, low cost, strength, compatability with many products.

Lead, the first of the three metals to be used in large-scale manufacture of tubes, occurs in nature in association with other minerals and usually requires intermediate processing before smelting and refining. Among the many properties of this metal that make it ideal for metal tubes are resistance to corrosion, softness and pliability, and ability to alloy readily with certain other metals.

Tin, widely used as a collapsible tube metal in Europe in the 19th century, comes from Malaya. The advantages of tin in the production of metal tubes are malleability, resistance to corrosion, strength

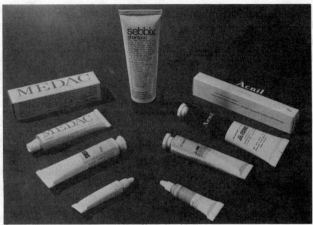

Courtesy of The Metal Box Co., Ltd.
FIG. 3.8. VARIOUS METAL TUBES

allied to lightness, and nontoxicity. Tin is generally recommended for pharmaceutical preparations that demand a container that is chemically inert. Tin is alloyed with a small amount of copper (0.5%) or other suitable stiffener to obtain the optimum degree of strength and collapsibility. With lead, antimony (usually 3%) is added to increase hardness. Tin and lead are combined to make two other types of tubes—tin-coated lead tubes and tin-lead alloy tubes. Normally the former consists of the proper amount of tin rolled on each side of the blank from which the tube is extruded. By weight the tin on the outside is generally 3% and on the inside 5%, but the percentages may vary.

Method of Manufacture.—Metal for tubes is usually melted from pigs, molded into slabs which are rolled out to proper thickness for the length, and punched into disc-shaped slugs of predetermined size. It is also possible with soft metals to use casting methods for the manufacture of slugs. After tumbling and lubrication the slugs are fed into presses and extruded under high pressure. These contain the tools necessary to provide the dimensional requirements. During the extrusion process cold plastic deformation of the metal is accomplished when the metal fills the die cavity and the excess flows up the straight sides of the cylindrical punch. The result is a one-piece, seamless, hollow tube with a shoulder (shoulder embossing if required) and neck at one end and open for the full diameter at the other end for filling. At this point the tube may or may not have an aperture in the neck. During the finishing operation, which is usually accomplished on a multispindle-type machine, the tube is trimmed to the desired length, the thread rolled on, and the front end of the neck faced so as to provide a square surface for sealing inside the cap. Aluminum tubes generally are annealed at this point in ovens at temperatures between 900° and 1200° F for a relatively short period.

Decoration of Tubes.—If tubes are to be printed (and the bulk of them are) a base coating of enamel is applied and the tubes pass through drying ovens. While the coating is not fully dry, the decorative inks and printing are applied on a multispindle offset printing machine. As many as four colors may be applied exclusive of the base coat, but each color involves the use of a separate plate. Occasionally an external coat of clear lacquer, which enhances the luster of the lithography, is added as protection against products that are known to attack inks. Internal coatings of waxes are sometimes necessary to prevent contamination of the product by the metal and in certain cases of product attack on the metal. This is of major importance because of the extremely broad range of products packaged in collapsible metal tubes. There are three main types of internal coatings: various waxes, unconverted interior lacquers, and thermosetting linings. The latter are converted resins, which are the toughest and most impervious of the linings in use, characterized by cohesiveness, continuity and flexibility of film.

BIBLIOGRAPHY

ANON. 1973A. Metal containers and the environment. Report of the working party on the design, use and disposal of metal containers. British Tin Box Manufacturers, pp. 1–76.
ANON. 1973B. Specification for cold-reduced tinplate and cold-reduced blackplate. Brit. Std. pp. 1–27.
ANON. 1973C. A-Z of the tinplate can. Tin Intern. *46*, No. 9, 322–323.
ANON. 1973D. Beer cans to be filled at 1200 a minute. Packaging News, p. 15.
ANON. 1973E. Easy-open tinplate can. Tin Intern. *46*, No. 10, 369.
ANON. 1973F. New progress in metal packages. IDE *14*, No. 170, 3–4. (Spanish)
ANON. 1973G. Warning on canned foods. Tin Intern. *46*, No. 11, 406.
ANON. 1973H. Lacquered cans and food compatability. Tin Intern. *46*, No. 11, 401.
ANON. 1973I. Stainless steel strip makes pressure labels. Packaging News, p. 18.
ANON. 1974A. Tin-free steel for can making. Tin Intern. *47*, No. 3, 81–82.
ANON. 1974B. A-Z of the tinplate can. Tin Intern. *47*, No. 4, 112–113.
ANON. 1974C. Tinplate packaging problems and prospects. Packaging India *6*, No. 3, 50–57.
ANON. 1974D. Technological aspects of tinplate. Packaging India *6*, No. 3, 58–70.
ANON. 1974E. Captive can making. Tin Intern. *47*, No. 5, 149–150.
ANON. 1974F. Production rate rises for cans of all sizes. Packaging News, p.16.
BARRY, B. T. K., and EVANS, C. J. 1974. D and I cans—a can-making revolution. Tin Uses No. 100, 12–15.
CLAUSEN, E. 1974. The environmental impact of metal can manufacture. Aluminum Intern. No. 9, 3.

GROB, G. 1973. International standards steel drums: Time for a re-think? Mech. Handling *60*, No. 9, 17–18.

SLOAN, J. 1973. Bibliography on double reduced tinplate. British Steel Corp. *SM-BIB-901*, 1–23.

SPAAS, J. 1972. Problems of limiting global migration of some non-ferrous metals. Ann. 1st Super. Sanita *8*, No. 2, 512–520.

SUMNER, J. 1973. Aluminum attractive for aerosols but steel cans cheaper. Engineer *237*, No. 6138, 50.

CHAPTER 4

Plastics

INTRODUCTION

Although it might seem that plastics materials are relatively new, the first of the synthetic resins was discovered in 1868. Plastics did not attain industrial status in the United States until the late 1930's and early 1940's. Today the plastics industry ranks as one of the billion dollar industries in the United States. Until the 1973-1974 petroleum crisis it was one of the fastest growing—showing an increase of more than 200% during the past ten years. The growth and development of the plastics industry has been so rapid that public acceptance of it and its products has far outstripped public understanding. This is unfortunate since a prerequisite to study any plastics material is an understanding of the industry of which that material is a part. As estimated by the Society of the Plastics Industry (SPI), the U.S. plastics business has at least tripled its investments in raw materials since 1946. It is presently spending millions of dollars each year in research alone. Today nearly 6,000 companies engage in some phase of plastics processing, turning out an annual volume in excess of seven and one-half billion pounds.

First Discoveries

In 1868 a manufacturer of billiard balls was looking for a material to replace ivory, which was difficult to obtain, expensive, and not really well-suited for making billiard balls. Seeking to enlist the aid of others, the manufacturer offered a $10,000 prize to anyone who could come up with a suitable material. John Hyatt discovered a cellulose base material which, when molded under heat and pressure, became plastic-like and could be made into a smooth, hard ball. Thus celluloid, the first synthetic plastic, was born.

Celluloid.—Celluloid or cellulose nitrate is not used to any great extent today, but for 41 years it was the only man-made material of commercial significance. After its initial application it was quick to find use in other areas. Celluloid eyeglass frames, combs, brush handles, toys, piano keys, and many other articles were developed and marketed. The collars and cuffs worn in the Gay Nineties were made of celluloid which could be wiped clean with a damp cloth. Later motorists in horseless carriages were protected by curtains of transparent celluloid, which could be rolled up and stuffed under the back seat of the car. Celluloid was highly inflammable, it became brittle with age, and tended to turn yellow and lose its transparency. However, it served well until the advent of the glass-enclosed car. With glass windows, however, another problem arose—glass shattered and created a safety hazard. The problem was solved by sandwiching a sheet of transparent celluloid between two sheets of glass and laminating the materials under heat. "Safety Glass" might break on impact, but the strong flexible plastic interlayer held the majority of the fragments in place.

By the turn of the century another industry found cellulose indispensible. Without a transparent flexible film motion pictures would have been out of the question. Celluloid served this purpose well for many years, but the fireproof films used today give evidence of its replacement by new, more suitable plastics.

Bakelite.—In 1909 Dr. Leo Baekeland discovered another synthetic plastic, which proved to possess many qualities not present in celluloid. "Bakelite," as this phenol-formaldehyde resin soon came to be called, could be formed into useful articles after being plasticized by the application of heat and pressure. Once molded Bakelite became a hard, insoluble mass. It would not soften (even if reheated) as celluloid would. The speed with which Bakelite was applied is explained by the fact that it offered properties numerous industries could not do without. The infant electrical industry, for example, needed suitable insulating material. Bakelite could be formed into flat sheets or solid objects of the most intricate shapes. It could be sawed, drilled, turned, punched and otherwise machined much the same as wood or metals. Telephone parts, distributor caps and other auto ignition parts, switch panels, instrument housings, machine tool parts, lighting fixtures, office equipment, composition board, etc. are examples of its many applications. Unlike celluloid, Bakelite remains very much a contender in the plastics industry today.

Recent Growth and Development

Dr. Baekeland's discovery was important for at least two reasons: first, it gave the industry a useful new plastics resin; second, and especially significant, it stirred the creative imagination of many people, suggesting new possibilities inherent to synthetic materials. From the time of Dr. Baekeland's discovery until the mid-1930's, development and growth within the infant plastics industry began to attract attention.

The Newer Materials

Cellophane was discovered by the Swiss chemist, Brandenberger, in the early 20th century and became commercial in the late 1920's.

Cellulose acetate was the next large volume plastic material to be developed commercially in the United States. Launched in 1927 it was available only in the form of sheet, rod and tube. In 1929 it appeared as a molding material and became the first injection-molded plastic.

Courtesy of Imperial Chemical Industries, Ltd.
Chemical Division, U.K.

FIG. 4.1. PVC BOTTLES

For the blow molding of PVC bottles and containers, ICI offers Welvic powder blends and compounds based on the specialist packaging polymer Corvic D50/16. Bottles made from Corvic and Welvic materials are shatterproof and resist the knocks and normal stresses which occur on the filling line, during transit, and in storage. Their lightness minimized transport costs. Many attractive bottles are made from Corvic and Welvic for use in garden and garage, kitchen, bathroom and bedroom.

The first of the vinyls, polyvinyl chloride, was discovered in 1927 but was not commercially launched until 1936. The vinyls are at present among the most important of all synthetic resins, making up a large family of materials. Polyvinyl chloride and its many derivatives are today considered vitally important in the plastics industry.

Early experimentation in plastics was mostly hit-or-miss and not until the mid-1930's and the work of Wallace Hume Carothers did the industry really begin to stand on a scientifically-oriented foundation. Carothers formulated theories regarding the atomic structures of plastics, resins, fibers and rubbers which proved to be the key to controlled experiments. The development of nylon (one of the most widely known of all plastics materials) was a direct result of his work, and the synthetic rubber program of World War II drew much of its impetus from his theories.

Plastics were quickly put on the critical materials list in World War II and their development accelerated rapidly. Both polyethylene and the polyesters were introduced in 1942. Later came the acrylics and the urethanes. Unfortunately the public heard little about the new plastics, mainly because they involved essential production. When their many war-proven abilities began to find expression in civilian products, plastics were found in unexpected places doing surprising jobs.

Definition of Plastics

Based on the discussion of celluloid and Bakelite one may define plastics as: "Synthetic or man-made materials produced by the application of heat and pressure and which can be formed into useful objects." Another definition is: "Any one of a large and varied group of man-made materials consisting wholly or in part of combinations of carbon with water, oxygen, hydrogen, nitrogen, and other organic and inorganic elements which, while solid in the finished state, is at some stage of its manufacture made liquid, and is thus capable of being formed into various shapes, most usually through the application, either singly or in combination, of heat and pressure."

The essential constituent of any plastics material is its resin or binder. The resin determines the physical and chemical characteristics of the various members of the plastics family. The resin is the basic plastic ingredient. Just as cement holds together fillers such as sand, gravel and rock to form a permanent solid, a plastics resin binds fillers or organic and inorganic substances together in a desired shape. Like cement plastics resin becomes "plastic" or moldable at some particular stage of its manufacture and becomes a solid mass at a later stage. Resins are usually prepared in the form of powder, flakes, granules, and, for some purposes, liquids. They are classified according to whether or not they can or cannot be made reformable by repeated application of heat.

Resin Constituents.—Resins are made up of elemental constituents. Nature has provided over 90 basic elements, each of which by definition is chemically indivisible. These elements exist in the form of solids, liquids, and gases, some of the more common being carbon, oxygen, hydrogen and nitrogen. Elements combine or compound with one another according to fixed ratios, because every element possesses a fixed number of bonds by which it can unite with certain other elements. Carbon has four bonds, oxygen has two and hydrogen has one:

$$-\overset{|}{\underset{|}{C}}- \qquad\qquad -O- \qquad\qquad H-$$

Various combinations of these elements form simple compounds:

$$H-\overset{\overset{\textstyle H}{|}}{\underset{\underset{\textstyle H}{|}}{C}}-H \qquad\qquad O=C=O \qquad\qquad H-\overset{\overset{\textstyle H}{|}}{\underset{\underset{\textstyle H}{|}}{C}}-\overset{\overset{\textstyle H}{|}}{\underset{\underset{\textstyle H}{|}}{C}}-H$$

CH₄ methane $\qquad\qquad\qquad\qquad\qquad\qquad\qquad$ C₂H₆ ethane

CO₂ carbon dioxide

Notice that none of these elements combines in a way that calls for use of more than its allotted number of bonds.

Courtesy of The Metal Box Co., Ltd.

FIG. 4.2. SPICE IN PRINTED PLASTIC PACK
Wright Crossley and Co., Ltd. are packaging their Lion brand of spices in
The Metal Box Co., Ltd. 1-oz standard plastic spice pack.

Plastics resins are made up of long chains of elements which have chemically linked themselves to one another. In plastics terminology, a combination of elements which are linked together as a unit and which can link with other elements is called a monomer. When this identical unit combination is repeated many times and the monomers are joined together in a chain-like structure, a polymer results. Plastics are referred to as high polymers when a unit monomer is repeated many times to form the polymer molecule. A long chain polymer is the building block from which plastics resins are made. If, for example, A- represents a monomer, a high polymeric structure might be represented as: A-A-A-A-A-A-A-A-A-A-A-, etc., with the single unit monomer, A-, being repeated from 500 to 1000 times. Sometimes a polymer may be made up of unlike monomers arranged in repeating patterns. A- being one monomer and B- being another of differing characteristics might be joined as: A-A-B-B-B-A-A-B-B-B-A-A-B-B-B-A-A-B-B-B-, and so on. Such a combination of differing unit monomers is called a copolymer. The long linking process within a plastics resin is known as polymerization, otherwise defined as the reaction by which single molecules (monomers) are linked together to form large chain-like structures (polymers or copolymers). For example, vinyl chloride monomers link together to form polyvinyl chloride.

Specific Gravity.—Specific gravity refers to the weight of a given volume of a substance as it compares to the weight of an equal volume of water. If 1 cu ft of water weighs 62.4 lb and 1 cu ft weighs 124.8 lb, the specific gravity is the ratio of plastic to water, or 124.8:62.4. In this oversimplified case the resulting specific gravity of plastic is 2. A plastics resin which is made up of relatively heavy molecules linked in a closely packed chain will have a greater weight per unit volume (specific gravity) than a resin whose molecular structure is made up of lighter, less closely packed chains. By knowing the specific gravity of a plastics material we can determine how much space a specified weight of that material will occupy. Subsequently we can determine the volume or amount of material which that particular weight will "yield." In other words, specific gravity indicates yield per pound of material.

Types of Plastic Materials

Plastics resins are usually divided into two broad categories: thermosetting and thermoplastic. Bakelite was the first of the thermosetting resins. Cellulose nitrate (or celluloid) was the first of the thermoplastics. When a thermosetting resin is heated if softens but stays soft only for a short time. Subjected to continued heat it cures or becomes hard. This curing represents a permanent chemical change. If the plastic is now reheated, even to the point of charring or burning, it will not soften. If a thermoplastic resin is heated to the proper temperature, however, it will become soft and plastic and not become hard again until cooled. This cycle of heating to soften and colling to cure may be repeated time and time again. Paraffin can be described as a thermoplastic.

Compounding Plastics.—Compounding is a process in which basic resins and other materials are intimately mixed together in as nearly a homogeneous mass as possible prior to subsequent processing into finished materials. Because of the complex nature of the various materials involved, compounding is one of the most difficult of all plastics industry operations and is recognized by many as a highly technical art.

Just as the resin material may vary considerably as to chemical makeup, molecular weight and configuration, and particle size, the types of plastics compounds may range from an adhesive (such as to bind plywood) to a coating solution (for papers) to a fine powder (for molding, extruding or casting into useful objects). Some plastics compounds are quite simple in composition, containing from 90 to 95% of the basic resin and only small amounts of additives. This is usually the case of clear plastic films. Other plastics are quite complicated by the addition of as many as a dozen ingredients mixed with as little as 20 to 30% of the basic resin. These are referred to as filled plastics from which many molded articles are formed. The majority of secondary materials or ingredients compounded with the basic resins are: (1) fillers, (2) plasticizers, (3) colorants, and (4) other miscellaneous additives.

Fillers.—The variety of fillers include inorganic, organic, mineral, metal, and synthetic materials. Fillers are more commonly used with the thermosetting resins, although there are also some filled thermoplastics. Where large amounts of fillers are used they are commonly referred to as extenders because they increase or extend the bulkiness of the material and likewise reduce its basic cost. A typical filler material is the fiberglass cloth used with hard-drying resins in boat building. In addition to extending the bulkiness of plastics articles, some of the more common functions of fillers are reinforcement, hardness, thermal insulation, chemical resistance, and enhanced appearance.

Plasticizers.—For a plastics material to exhibit flexibility, resiliency and flowability it must be plasticized, either within itself or through the addition of a substance known as a plasticizer. Without this additive it would not be possible to make films, sheeting, tubing and other flexible forms of plastics. The theory of plasticization concerns the internal movement of the molecular chains within

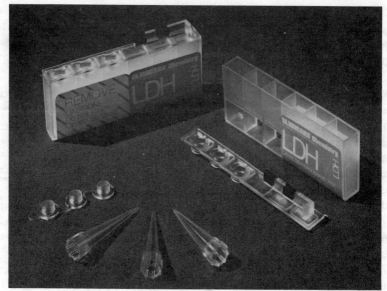

Courtesy of Imperial Chemical Industries, Ltd.
Chemical Division, U.K.

FIG. 4.3. TPX IN USE

The kit is disposable and is made from a TPX polymer manufactured by ICI. The cuvette is subjected to several critical processes and must, therefore, be made to exacting specifications. These include: chemical compatability with blood and all the chemical reagents used in the cuvettes; transparency to visible and ultra-violet light of precise selected wavelengths; resistance to heat at incubation temperatures of up to 90°C; ease of molding and the ability to be sealed ultrasonically. Only a plastics material is practical for this application and, of the transparent plastics commercially available, only TPX polymer meets all the requirements.

polymeric structures. The plasticizer allows these molecular chains to move with respect to one another with a minimum of entanglement or friction, thus acting as an internal lubricant to overcome the attracting forces between the chains and to separate them to prevent intermeshing. The higher the temperature in processing, the greater the penetration of the plasticizer in between the chains and consequently the greater the actual plasticity‚ or flexibility of the material.

Most plasticizers are liquids which exhibit good compatibility with resins. Plasticizers are usually colorless and have low vapor pressures and good thermal stability. Added to the basic resin, with or without other ingredients, the plasticizer imparts increased flexibility, impact resistance, resiliency, moldability and softness. However, plasticizers may decrease strength characteristics, heat resistance, dimensional stability and solvent resistance. Therefore the type and degree of plasticizer used is greatly determined by the characteristics desired in the finished material.

Colorants.—Color is an important characteristic in plastics. Colorants currently used make possible colored materials varying from pastels to deep hues as well as vari-colors and marble-like shades. Two primary types of colorants are presently used in plastics: dyes and pigments. The essential difference between them is solubility. Dyes are fairly soluble in plastics compounds while pigments, being relatively insoluble, are dispersed throughout the plastics mass. Pigments are more stable than dyes and tend to stand up better throughout the life of the material. A recent development in plastics coloring is the color concentrate which is furnished with plastics compounds by compound processors. It enables the materials processor to color his plastics by mixing some of the color with the plastics ingredients or by adding it directly to the materials as they are being processed.

Miscellaneous Ingredients.—Other ingredients used in compounding plastics in small quantities, usually ranging from 0.5 to 2.0% of the total composition, include: stabilizers, inhibitors, catalysts, hardeners, and lubricants.

Because many plastics are organic in nature they are subject to chemical attack by the elements of the air, which can give rise to degradation or aging. Since much of this attack is in the form of oxidation, certain ingredients may be added to a plastics compound to effect a more stable temper to its physical and chemical properties. Such additives are called stabilizers and are used in compounds to lengthen shelf-life prior to processing as well as to maintain the characteristics and properties of the finished product. Inhibitors may be classified as temporary stabilizers as they are used to lengthen the shelf-life of in-storage compounds. In some instances the plastics compound may contain monomeric or low polymeric ingredients which tend to polymerize over long periods of time while in-storage (even at room temperature). The amount and nature of the inhibitor is usually such that at the higher processing temperatures, it is either destroyed or becomes so ineffective that it does not interfere with polymerization or curing during the processing cycle.

Also important in the final polymerization of plastics are catalysts, which are added in small quantities to aid in curing.

For hardness in plastics various reactants or hardeners may be added to a compound. It is important to differentiate between hardeners and plasticizers. Plasticizers are used to regulate rigidity, not hardness. The rigidity of a material is a measure of its flexibility. Hardness refers to the ability of a material to resist deformation or penetration by an outside force.

Finally, in the case of compounds which will be molded into finished objects, lubricating additives are essential. They act as a basting agent which facilitates the removal of the plastics article from the mold.

Chemistry of Plastics

Plastics used for the manufacture of packaging materials are generally composed of a polymer or resin to which a number of additives are added before the molding takes place. The term resin invariably means the pure polymer. The resin itself can be produced by polymerization, which consists of polycondensation and polyaddition.

In addition polymerization, the monomers, which must obtain a double bond, form chains or macromolecules in the presence of a catalyst or initiator.

$$n \left(\begin{array}{cc} R & R \\ | & | \\ C = C \\ | & | \\ R & R' \end{array} \right) \qquad R^1 \left(\begin{array}{cc} R & R \\ | & | \\ C - C \\ | & | \\ R & R' \end{array} \right)_n -R^1$$

R^1 is part of the molecule of the catalyst, while R varies in the different polymerization products. Examples of addition polymerization products are polyethylene and polypropylene. When more than one type of monomer is joined together the process is known as copolymerization.

Polycondensation is the formation of chain macromolecules by a chemical reaction of two different monomers at two sites in the molecule with the loss of water, or another product.

$$R - C \overset{OH}{\underset{O}{\diagdown}} \ + \ R' - NH_2 \longrightarrow \left[R - \underset{\underset{O}{\|}}{C} - \underset{|}{\overset{H}{N}} - R' \right]_x + H_2O$$

Polyamides, polyesters and phenol formaldehyde resins are polycondensation products.

TABLE 4.1

PROPERTIES OF PLASTICS TO BE CONSIDERED IN THEIR SELECTION

Advantages	Disadvantages
Fabrication—Most plastics are readily adaptable to mass production methods, and intricate products of close dimensional tolerances can be made, produced in large quantities and at low manufacturing costs.	Repair—Broken plastics are seldom mendable, and it is usually better to replace them.
Weight—Generally the density of plastics is less than that of most metals; they offer great weight savings.	Strength—As a rule plastics are not very strong, even when allowances are made for their low density.
Corrosion—Most plastics are immune to rust, rot or corrosion.	Stability—Plastics may warp, shrink or "creep" and they are relatively soft and easily scratched.
Insular Qualities—Plastics are characteristically poor conductors of heat and electricity.	Thermal Qualities—Plastics cannot be operated at red heat; most can be burned; some are fire hazards.
Color—Transparent products and products of an unlimited color range are available in plastics.	Odor—Many plastics have definite odors, not all of them pleasant.
Cost—While most plastics lend themselves to relatively inexpensive manufacturing operations, they themselves are not cheap. Satisfactory plastics materials may be quite expensive, especially in small quantities.	

Plastics Manufacture

The manufacture of plastics products can be said to consist of three basic steps: the first consists of isolating the basic resin materials by processing a polymeric material from various chemical compounds. In the second step the resin is mixed and compounded with other materials to produce an intermediate material compound which is ready for processing. Finally plastics compounds are processed into useful finished materials by the use of heat and pressure.

Plastics Processing Methods

The variety of techniques presently used in the processing of plastics materials is almost as diverse as the number of different resins which make up the plastics family. One of the most common methods for producing rigid plastics products is molding. It can produce articles ranging from hard, rigid items such as ash trays, instrument panels and intricate parts to flexible and semirigid objects

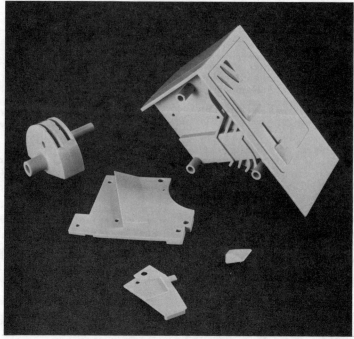

Courtesy of E. I. DuPont De Nemours and Co.

FIG. 4.4. SMOOTH, HARD SURFACE OF ACETAL MOLDING RE-
DUCES WEAR AND ELIMINATES COIN-JAMMING PROBLEMS IN
THIS COIN PLATE ASSEMBLY WHICH IS SNAP-FITTED TOGETHER.

such as detergent bottles, toys and synthetic rubbers. Other methods
include reinforcing, laminating, casting, extruding, calendering and
coating.

Compression Molding.—Compression molding is the most common
of the molding methods. Thermosetting resins and fillers, usually in
the form of powder or granules, are placed directly into the mold
cavity. The mold is closed and heated, pressing down upon the raw
materials and causing them to flow throughout its shape. While the
heated mold is closed, the thermosetting materials undergo a
chemical change which permanently hardens them into the shape of
the mold. After cooling the molded plastics object is removed from
the mold.

Transfer Molding.—Transfer molding is similar to compression
molding in that the plastic compound is cured under heat and
pressure. It differs in that the plastic is heated to the point of
plasticity before it reaches the mold and is hydraulically forced into
it. This process was developed especially for molding small intricate
parts with small deep holes or numerous metal inserts, as it allows

the liquified plastics material to flow around these metal parts without causing them to shift position.

Thermoforming.—A thermoformed plastic tray is produced by heating a plastic sheet to softness. By a variety of techniques it is formed to a desired shape and allowed to cool. The sheet is then trimmed and removed from the mold. Several different methods are used to produce thermoformed trays, cups or tubs (i.e., pressure, vacuum forming and/or drape methods). The selection depends on the specific plastic sheet used and machinery available. Food processors may purchase formed trays from suppliers or may form in-line with filling and closing. Many food processors have totally integrated in-line operations in their plants, because the economics are quite favorable in high volumes.

Injection Molding.—Injection molding is the principal method for forming thermoplastic materials, although modifications of it may be used with thermosetting materials in some special cases. A plastics material, usually in powder or granulated form, is loaded into a hopper which feeds it to a heating chamber where a plunger pushes it through to a nozzle. The fluid plastic is forced, by pressure of the plunger, through the nozzle and into a chilled mold where it hardens. The mold is subsequently opened for its removal.

Blow Molding.—Blow molding, which is used for thermoplastics, consists of stretching and hardening a plastics material against a mold

Courtesy of Uniloy, Inc.

FIG. 4.5. BLOWMOLDER

This photo illustrates a 350R Uniloy Blowmolder for fabricating plastic half-gallon containers.

by air pressure. There are two primary methods of blow molding. In extrusion blow molding, a gob of molten resinous material is formed into a parison or shape which approximates the shape of the desired product. The parison is inserted into a female mold where air pressure expands it, like a balloon, against the mold walls there it is cooled and removed.

In injection blow molding a plastic form previously made by injection molding is clamped between a die and cover. Air pressure forces the material into contact with the mold, which is contoured to match the desired shape of the finished product. The plastics article is subsequently cooled and removed from the mold. Injection blow molding is increasingly used to obtain precision finishes on bottles.

Laminating.—Laminated packaging products such as foil/paper/film combinations are commonly known. However, lam-

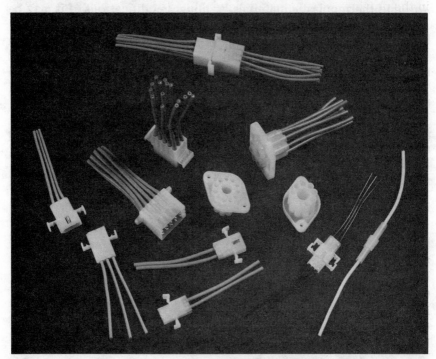

Courtesy of E. I. DuPont De Nemours and Co.

FIG. 4.6. MULTIPIN ELECTRICAL CONNECTORS MOLDED FROM NYLON ARE STRONG, RESISTANT TO CHEMICALS, HIGH IN DIELECTRIC STRENGTH, AND SELF-EXTINGUISHING

Latter feature is important safety factor in case of fire or short circuit equipment.

inated plastics encompass a far greater range of application. Many materials such as glass, cloth, paper or woven glass fibers may be impregnated with a resin, layered, and then cured by heat and pressure to produce a laminated plastics material.

High-pressure Laminating.—The terms high pressure and low pressure (or reinforced) are commonly used in the plastics industry, but the distinction between them is not always clear. When a pressure of more than 1,200 psi is used the product is generally called a high-pressure laminate. The production of such laminates usually involves stacking the desired number of resin-impregnated sheets on a press, and then applying heat and pressure to bond the sheets and cure the resin. Because a press of relatively high pressures will be required, such laminates will generally be limited to flat sheets and simple shapes, such as corrugated panels. Rods and tubes and other simple cross sections are easily produced by high-pressure laminating, but more complicated shapes and continuous lengths fall in the province of low-pressure laminating (or reinforcing).

Reinforcing.—In low pressure or reinforced processing pressures usually range from 50 to 350 psi. In some cases a process similar to blow molding may be used, in which impregnated materials are placed in a female mold and drawn down by air pressure or vacuum to shape the product. In other cases, such as in boat building, fibrous materials are laid over the mold surface, arranged to the desired thickness and smoothed down. A liquid resin is then applied to laminate and cure the materials into a solid mass. Where both male and female molds may be used, as in the production of one-piece boat hulls, the weight of one mold on the other may supply the necessary pressure.

Plastic Resins

Because of the many source materials at the disposal of the plastics chemist, and because so many different combinations of these materials are possible, the family of plastics today embraces a large number of members, each of which differs in properties and characteristics. Some resins are used in film form as well as in rigid parts. Due to the large number of resins this chapter will describe end applications of plastics resins used primarily in rigid and semirigid packaging.

Primary Plastics Materials

Listed are some of the more important plastics resins and some applications of each. This listing is by no means complete, neither by resin name nor diversity of application. (Capital letters in parentheses beside resin names indicate major plastics categories: (S) indicates thermosetting and (P) indicates thermoplastic.)

Courtesy of Montedison, Italy

FIG. 4.7. HI-DENSITY POLYETHYLENE
Moplen-RO high-density polyethylene crates for the harvesting and transport of fruit and vegetables.

ABS (acrylonitrile butadiene styrene) (S)—tubs, trays, thermoforms, closures.

Acrylics and acrylic multi-polymer (S)—base materials for treating fabrics and papers, automobile taillights, brush backs, windows for industrial plants, "piped" light devices, plastic bottles, closures, thermoforms.

Alkyds (S)—electronic parts, appliance handles, high- and low-pressure laminates, protective coatings.

Butadiene styrene (BDS) (P)—boxes, trays, tubes, toys, bowls.

Butyrates (cellulose acetate butyrate) (S)—gun stocks, furniture strips (rattan), lawnmower rollers, radio grilles, wallboard molding, tool handles, vials, bottles, thermoformed pieces.

Casein (S)—buttons, game counters, knitting needles, adhesive component.

Cellulose acetate (P)—business machine keys, combs, eyeglass frames, packaging film, table mats, shoe heels, vials, bottles, thermoformed pieces.

Cellulose nitrate (P)—Because this material is flammable, its use is restricted; it is, however, still used for piano keys, tools, handles, fountain pens and film coating.

Ethyl cellulose (S)—proximity fuses, refrigerator door strips, toilet seats, toys, flashlight cases.

Epoxies (S)—drop hammer dies, gymnasium floors, pipe linings, printed circuits, adhesives and coatings.

Fluorocarbons (polytetrafluoroethylene) (P)—valve seats, pump diaphragms, tier mold gaskets, wire coating, high-voltage insulation, coaxial cable spacers, radar parts.

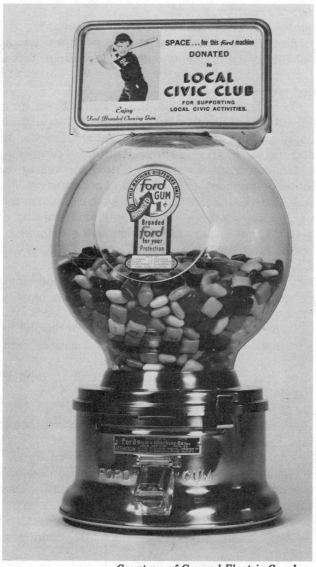

Courtesy of General Electric Co., Inc.

FIG. 4.8. POLYCARBONATE BUBBLE

This bubble dome is molded from polycarbonate resin.

Nitrile polymers (AN and SAN) (P)—plastic bottles, thermoformed
pieces.

Nylons (polyamides) (P)—slide fasteners, brush bristles, gears,
fishing lines, raincoats, surgical sutures, gears, tennis racket
strings, thermoformed pieces.

Polyallomers (P)—bottles, sheets.

Polycarbonates (P)—electrical switchboard connectors, coil forms, relay bases, safety helmets, diffusion lenses for hospital lamps, battery components, capacitor housings, fuse covers, hose couplings, sight glasses, rigid trays, bottles, thermoformed pieces.

Polyesters (S)—chief impregnant of mats of glass fibers for making reinforced plastics structures, pipes, luggage, protective coatings, washing machine parts, antenna masts, thermoformed pieces, packaging films. Reinforced plastics (a polyester-glass combination) are being used for ceiling canopies and other components in building and construction.

Polyethylene (P)—flexible ice trays, garbage cans, garden mulching sheets, laundry baskets, squeezable bottles, waste baskets, packaging films, paper coating, lining for swimming pools, rain protection for tennis courts, coatings for milk cartons, bottles.

Polystyrenes (P)—radio housings, refrigerator food containers, instrument panels, battery cases, wall tile; blow-molded, injection-molded and thermoformed pieces; packaging film.

Polypropylenes (P)—valves, packaging film, wire insulation, pipe and fittings, textile machinery parts, aerosol containers, plastic bottles.

Polyurethanes—rug underlays and many other foam products such as cushions, mattresses, insulated clothing.

Propionates (P)—vials, bottles, thermoforms.

Silicones (S)—switch parts, adhesives, glass cloth laminates, textile finishes, induction heating apparatus.

Ureas (S)—kitchen accessories, tableware, closures, brush backs, radio cabinets, also in resin form as plywood adhesives, baking enamel, coatings and paper and textile treatment, closures, boxes.

Vinylidene chloride (Saran) (P)—packaging films, sheeting, chemical piping, bristles, upholstery filaments, window screens, filter cloth.

Vinyls (vinyl chlorides) (P)—raincoats, upholstery, draperies, wall and floor tile, garden hose, wire insulation, phonograph records, shoe soles, billfolds, purses, luggage, umbrellas, inflatable toys, lamp shades, bottles, packaging film, thermoformed pieces.

BIBLIOGRAPHY

ANON. 1973A. New LDPE film. Mod. Packaging 46, No. 9, 116.
ANON. 1973B. Starch biodegrades plastics. Mod. Packaging 46, No. 10, 10.
ANON. 1974A. Anti-blocking agent. Europlastics 47, No. 1, 21.
ANON. 1974B. New materials round-up. Europlastics 47, No. 1, 48-50.
BARRETT, J. W. 1973. Economic incentives and constraints for using plastics. The Plastics Institute, London, 1-13.
BAUM, B., and DEANIN, R. D. 1973. Controlled UV degradation in plastics. Polym.-Plast. Technol. Eng. 2, No. 1, 1-28.
BAWN, C. E. H. 1973. Polymers: Developments for the future. J. Occa 56, No. 9, 423-429.
HALL, C. W. 1973. Permeability of plastics. Mod. Packaging 46, No. 11, 53-57.
HINSKEN, H. 1973. Plastic film wraps for pre-cooked foods. Australian Packaging 21, No. 9, 27-31,33.
INGLE, G. W. 1972. An industry view of plastics in the environment. Polymer Preprints 13, No. 2, 635-642.
KUSTER, E., and AZADI-BAKHSH, A. 1974. Studies on microbial degradation of plastics films. The Plastics Institute, London, Paper 16.
LUTZ, J. T. 1973. Acrylic modification of plasticized poly (vinyl chloride). Am. Chem. Soc. Org. Coatings Plastics 33, No. 2, 97-104.
OGORKIEWICZ, R. M. 1974. Thermoplastics: Properties and Design. John Wiley & Sons, New York.
OKADO, T. 1974. Hygienic problems of plastics. Japan. Plastics Age 12, No. 1, 44-49.
POTTS, J. E. GLENDINNING, R. A., and ACKART, N. B. 1974. The effects of chemical structure on the biodegradability of plastics. The Plastics Institute, London, Paper 12.
SCOTT, G. 1973. Light-degradable plastics. Australian Packaging 21, No. 11, 31-34.
WHITE, L. J. 1973. Whither degradable plastics? Chem. Eng. 80, No. 24, 68, 70.

CHAPTER 5

Paper, Paperboard and Corrugated Fiberboard

HISTORY

Paper manufacturing was invented by the Chinese around 150 A. D. The English established a paper industry in the 17th century and the first paper mill in the United States was built in 1690. Early paper production in England and the United States used rags as the raw material. The rags were ground into a pulp and diluted. A screen was dipped into the pulp and shaken. The pulp sheet was pressed to remove more water and then dried. A moving screen was later invented by the Frenchman, Louis Robert. This invention, the forerunner of the modern Fourdrinier machines, was improved upon and later sold to the Fourdrinier brothers. At almost the same time, cylinder paper-making machines were also developed. These two types of machines produced paper in such great volume that a shortage of rags for use as raw material developed. A new source of cellulose material had to be developed.

In 1841 Friedrich Keller of Germany invented a mechanical process for making pulp from wood. This was followed by a chemical process called the soda process invented in 1851 by two Englishmen, Hugh Burgess and Charles Watt. In 1866 an American, Benjamin Tilghman, invented the sulfite process, followed by Dahl's sulfate process in Germany in 1889. In 1880 the sulfite process was modified into an indirect process by the German, Mitscherlich, and into direct steam injection by two Austrians, Ritter and Kellner, in 1882. Sulfite was the major U.S. chemical pulping technique until 1937. Today, however, the sulfate process dominates in this country.

Pulp produced by the mechanical process contains all wood except the bark. Pulp produced by chemical processes, however, is essentially cellulose—the unwanted and unstable lignin and other noncellulosic components having been dissolved and eliminated by the chemical treatment. Therefore, chemical pulps are superior to mechanical or ground wood-pulp for fine papermaking. However, chemical pulps are too expensive for the cheapest grades of paper such as newsprint.

DEFINITIONS

Paper is a generic term encompassing a broad spectrum of materials derived from cellulosic fibers. It is the oldest packaging material and also the most versatile. Paper (the generic term) refers to all types of matted or felted sheets of vegetable fiber formed on a fine wire screen from a water suspension and bound together by material weaving of the fibers and by bonding agents. In general the term paper refers to a lighter, thinner and usually more flexible material. The distinction between paper and paperboard is not always very clear, but paperboard is heavier and more rigid than paper. Usually all sheets which are greater than 12 points in gauge (i.e., 0.012 in. thick) are classified as paperboard. Fiberboard is paperboard used for manufacture of corrugated containers or of solid fiberboard shipping containers. Solid fiberboard is a heavy multiply paperboard. The raw material for the manufacture of paper, paperboard and fiberboard is pulp, a mixture of fibers usually obtained by chemical or mechanical treatment of wood. Pulp consists of cellulose with various amounts of other materials such as hemicellulose and lignin.

Production Of Pulp

As the volume of paper pulp consumed increased wood became the principal source of cellulose. Cotton, linen rags, hemp, and textile wastes are also used as they are available. Trees of about 4 or 5 in. in diameter are selected and cut into logs about 6 ft in length. These logs are fed into a slasher, a circular saw which cuts the logs into 24-in. to 48-in. lengths. These short logs are then debarked by means of rotating drums which literally rub off the bark, or by knife barkers which shave the bark off by means of rotating knives.

Mechanical Pulp.—After debarking, soft woods such as spruce and balsam are ground in water by means of burred grindstones. The ground pulp is floated away by means of a water spray. The ground pulp is screened and the oversized pieces are ground once again. The slurry is then thickened by screening. The overflow from the thickeners is called white water which is used to cool the grindstones. Mechanical pulp is used primarily in the production of cheaper grades of paper and paperboard. In the manufacture of newsprint, wall, tissue, and certain wrapping papers, the mechanical pulp is usually mixed with a small amount of chemical pulp.

Chemical Processes.—After debarking, the wood is chipped or ground into small particles. These finely-divided particles are fed into a digester where they are subjected to steam pressure in the presence of sodium sulfide and caustic soda (or sodium hydroxide in the case of the sulfate process) or other chemicals demanded by the process selected. The purpose of the chemicals is to separate the lignin and other undesirable chemicals from the cellulose. The sulfate or Kraft

Courtesy of St. Regis Paper Co.

FIG. 5.1. PULP MANUFACTURE

Mountains of wood chips in a space-age setting in a Monticello, Miss. kraft mill. The forest resources industry has reshaped the face of the south's economy since World War II and today ranks near the top in dollar value. This $116-million St. Regis mill produces in excess of 1535 tons daily of paper and paperboard.

process is capable of handling practically all types of wood. However, coniferous woods are used almost exclusively, since the process was developed to remove the large amounts of oil and resins in these woods. After digestion the pulp is screened, washed, and bleached. Some by-products, including tall oil, are recovered. Other chemicals are recycled. Kraft is the German word for "strong." Pulp from the sulfate or Kraft process is used in paper bags and other high-strength applications. Some Kraft pulp is often mixed with other pulps to increase their strength.

The soda process is very similar to the sulfate process. It is usually used on poplar, birch, maple, chestnut, and gum. The active chemical employed in digestion is caustic soda. The remainder of the processing steps are similar to those employed in the sulfate process. The woods used in the soda process yield pulps of shorter fibers (about 1-1/2 mm) than those from the conifers used in the sulfate process. Therefore, the soda process pulps are weaker and are usually mixed with Kraft pulp to fill in the spaces between the longer (2 to 3 mm) fibers. Most of this pulp goes into the manufacture of books, magazines, and tissue papers.

The sulfite and neutral sulfite processes rank second to the sulfate process in the quantity of pulp produced. Hemlock and balsam are usually selected. The liquor fed to the digesters is a solution of calcium and magnesium bisulfite. While the chemistry is entirely different from that of the other chemical processes, the end objective is the same—to separate the cellulose from the other components of the wood. Sulfite pulp is the grade used for the production of the finest papers, including the bond office papers. Sulfite pulp is used either alone or with some rag pulp to make writing and high-grade book paper.

Rag pulp is derived largely from textile mill scraps. The rags are separated by means of color. Rags are disintegrated into small pieces, then digested into caustic lime, caustic soda, or a mixture of caustic lime and sodium carbonate or soda ash. After digesting at 120°C for 10 to 12 hr the resulting pulp is screened out, washed, bleached, rewashed, and then made ready for manufacture into fine writing paper.

Manufacture of Paper

Pulps, even though frequently manufactured into coarse sheets, still lack the properties desirable in finished paper such as proper surface, opacity, strength, and feel. Pulp is converted into paper by two general processes: beating and refining. There is no sharp distinction between these two operations. Mills frequently use both processes since refining is fundamentally beating.

The most commonly used type of beater is also known as a Hollander. Pulp is disintegrated in the beater to make the paper stronger, more uniform, more dense, more opaque, and less porous. Fillers, coloring agents, and sizing are usually added at the beaters. The Jordan engine is the standard refiner. The Jordan is a rotating mill with steel bars permanently affixed to the lining. As the mill rotates, these bars beat the pulp for the desired effect. Filler, sizing or water resisting chemicals, and coloring may be added at the Jordan. The order in which the various materials is added is usually as follows: (1) Various pulps are blended to give the desired density and uniformity. (2) Filler is added with or just after the fiber. (3) After sufficient beating, the sizing is put in and mixed thoroughly. (4) Color is added. (5) Alum ($Al_2(SO_4)_3$) is introduced to produce coagulation and the desired coating of the fibers and to set the size and dyes.

Courtesy of St. Regis Paper Co.

FIG. 5.2. PAPER PRODUCTION—THE PAR FOUR MACHINE ROOM

Stretching more than 1,000 ft from the camera is this view of the area in which St. Regis makes paper and paperboard at its new Ferguson Mill near Monticello, Miss. Giant 1,130-ton-a-day paperboard machine is shown on the right and the 405-ton paper machine on the left. Total length of machine room is 340 yd, which would be a Par 4 hole on a championship golf course.

Most paper except absorbent tissues and blotting papers must have a filler. The purpose of the filler is to occupy the spaces between the fibers, thus giving a smoother surface, increased printability and whiteness, and improved opacity. Fillers are usually inorganic substances such as clays, precipitated calcium carbonate, titanium dioxide, blanc fixe (precipitated barium sulfate), Zeolex (precipitated hydrated sodium silico-aluminates and sodium calcium silico-aluminates).

Sizing is added to impart resistance to penetration by liquids. All papers other than absorbent papers are sized. Sizing may either be added at the beaters or applied to the paper surface after the sheet is formed. When it is added at the beaters, alum is later added to precipitate the sizing as a gelatinous film on the fiber to produce a hardened surface upon the loss of water of hydration. Sizing may also be applied upon the dried paper. In this process the sizing itself must have adhesive properties. Animal glue, gelatin, and starch are frequently used. Coloring may be applied either to the wet pulp or on the papermaking machine itself.

After the pulp has been treated, it is ready to be made into paper sheet largely on two types of machines: the Fourdrinier and the cylinder machine. The basic principle of operation is essentially the same for both machines. A suspension of pulp is sent to the machines where it is fed onto a moving, endless wire screen. It is shaken while on the screen to orient the fibers and also dewatered. It is carried under a "dandy" roll which smooths the surface. The sheets are then further dewatered by draining, rolled, and sent through drying rolls and finally wound on reels for ultimate use. The cylinder machine is employed in the manufacture of heavy paper, cardboard and paperboard in which several dissimilar layers are united into one heavy sheet.

Paper Coating.—Some papers demand a higher finish than would come directly off the Fourdrinier machines and so such papers are coated. Many processes, either on-machine or off-machine, can be used for coating paper. Paper exits the Fourdrinier screens, is partially dewatered, and is then sent over drying rolls. One method of coating is to insert a roll coated in between the several drying rolls. Thus a paper sheet passes over some drying rolls, through the paper-coating machine, and then through the remaining drying rolls. Various formulas are used in the coating mixture, but they usually consist of a mixture of products such as high-grade clay, Zeolex, precipitated calcium carbonate, or titanium dioxide with a binder such as starch. Off-machine coating would employ the same general process, but the starting paper would have been completely dry and another mechanical coating technique would be employed.

Board Types

Board types are generally designated by caliper (thickness), which is usually expressed in thousandths of an inch, with each one-thousandth (.001 in.) being called a point. Most boards are designated by caliper with the exception of Kraft linerboard stock which follows the paper industry pattern of specifying the weight per thousand square feet (lb per 1000 sq ft). There is no clear-cut dividing line between the heaviest grade of paper and the lightest board. For example, blotting stock is always referred to as paper although it may be as thick as 24 points. However, the lightest standard board is 7-1/2 points. Heavy papers may run 4 through 5 points (.004 to .005 in.) Boards may be classified as: (1) chemical pulp or virgin (usually) boards and (2) waste paperboards.

TABLE 5.1
PAPERBOARDS USED IN PACKAGING

Grade	Contents	Caliper Ranges (In.)	Characteristics	Uses
Plain chipboard (solid newsboard; filled news; news vat-lined chipboard)	100% low-grade waste papers	0.020–0.060	Lowest cost board produced; adaptable to special lining papers; not good for printing; color range from light gray to tan; poor bending qualities	In set-up boxes for candy, gifts, stationery, textiles, etc.
White wood, vat-lined chipboard	White liner is 100% groundwood; back is same as chipboard	0.020–0.060	The white liner is adaptable to color and can be sized for printing; poor bending qualities	For higher grade set-up boxes with white liners
Bending chip	100% waste papers of low grades	0.016–0.034	Generally the lowest cost board for folding boxes; usually gray or light tan, but can be printed with all colors; excellent bending qualities	Boxes for cigarettes, toiletries, pastry, suit cartons, etc.
Colored, manila-lined bending chip	Top liner is virgin pulps or high-grade waste paper, some groundwood; back liner same as bending chip	0.016–0.034	Same as bending chip, except for brighter liner; very even colors on liner; excellent bending qualities	Same as bending chip, but where more eye appeal is desired

Type	Composition	Thickness	Properties	Uses
Bleached, manila-lined bending chip	Top liner is high-grade waste virgin sulphite pulp, some groundwood; back is usually news or chip	0.016–0.034	White top liner permits special treatment and multicolor printing; excellent bending qualities	For products where eye appeal is desired, using 2 and 3 color printing
White patent-coated newsboard	Top liner is 100% virgin pulp and high-grade waste; free of groundwood	0.016–0.034	Smooth, very white board; colors not subject to fading; considerable strength; excellent bending qualities	Folding boxes, displays, posters, shirt and textile inserts, sleeves, etc.
Clay-coated board	Same as white patent-coated news-back except for clay-coated surface	0.016–0.024	Very smooth, white board; excellent printing surface; excellent bending	Wherever high-grade multi-color printing is needed
Solid bleached sulphate board (special food board)	100% virgin sulphate pulps	0.012–0.026	Good strength and performance on automatic filling machines; solid white; excellent bending	Food, bakery boards, ice cream boards, frozen food cartons, milk cartons
Unbleached solid sulphite board	100% virgin pulps	0.012–0.026	Less expensive than solid bleached sulphate; very clean; buff shade; excellent bending	Food, toys, hardware, mechanical parts—where eye appeal is desired
Solid manila board	Virgin pulps and high-grade waste substitutes	0.012–0.030	Available with white liner and manila back; excellent bending	Toys, hardware, mechanical parts

TABLE 5.1 (*continued*)

Grade	Contents	Caliper Ranges (In.)	Characteristics	Uses
Extra-strength Kraft boards	Kraft and Kraft waste pulps	0.016–0.050	Minimum burst and tear test values; brown or dark colors; excellent bending	Hardware, automotive parts, toys
Extra-strength white-lined Kraft boards	Kraft and Kraft waste pulps	0.016–0.050	Same as above, only top liner is white or a pastel color	Same as above where more eye appeal is required

Courtesy of West Virginia Pulp and Paper Co.
FIG. 5.3. INTERIOR VIEW OF BOARD MILL

Chemical Pulp Boards.—Chemical pulp boards are usually made of 100% Kraft fiber, either bleached, semibleached, or natural. Bleached boards, when processed without additives which might be dangerous or impart a disagreeable odor or taste to the board, are used for cartons which come into direct contact with food products. Semibleached boards and fully bleached boards are of similar quality except for brightness and color. Semibleached stocks have the familiar manila (buff) color usually seen in milk cartons.

Most chemical pulpboards are made on Fourdrinier machines. They are usually single ply, made form a single laydown of pulp on a moving wire screen. They may be double ply, in which case the second layer may be made by using specific fiber composition to provide certain special properties to the top of the sheet.

Fourdrinier boards tend to have less caliper variation across the web and a less pronounced grain direction than do cylinder boards. In the manufacture of paper or board the majority of the fibers tend to align themselves in a direction of travel through the machine. The direction taken is variously known as machine direction (MD), longitudinal or with the grain. It is at right angles to the cross direction (CD), sometimes referred to as across or transverse. The ratio of MD:CD stiffness is approximately 2:1 on Fourdrinier boards and 3:1 on cylinder boards.

The more uniform (squarer, in terms of directional stiffness) Fourdrinier sheet has proved desirable in many high-speed packaging operations. The cutting and creasing operation is, however, sometimes more difficult on solid Fourdrinier chemical boards because they are harder and there is no ply breakdown to assist the score. Solid chemical pulp boards are usually better benders since they do not contain the high percentage of short fibers found in waste or cylinder board sheets.

Natural Kraft boards (usually with paper-machine added wet strength) are used in the manufacture of many patented-type can and bottle carriers. These are used when maximum strength and low cost are desired. They are quite often machine-coated for printability. Cylinder boards are paper machine-coated or otherwise treated on machine to improve appearance and printing qualities. For example, white patent-coated boards are not actually coated. The term is used in the paper industry to designate board in which bleached pulp is used for the top liner.

Most standard Kraft boards range in caliper from 8 through 24 points (0.008 to 0.024 in.). Examples of boards available are: (1) bleached Kraft tag board; (2) bleached Kraft food-grade board (solid sulfate made from virgin pulp); (3) natural Kraft linerboard, suitable for high-strength folding cartons, beer carriers, etc.; (4) natural Kraft bottle carrier stock, with wet strength, usually natural interior and white exterior; (5) bleached Kraft natural Kraft back duplex board, used for heat-and-serve roll trays, and where a food-grade white interior is desired; (6) machine-coated bleached Kraft food-grade board, solid bleached sulfate or SBS made from virgin pulp, chlorine bleached; and (7) bleached Kraft carton board, with mold inhibitor, used for bar soap cartons.

Waste Paperboards.—These boards are all produced on cylinder paper machines, where the board is built up by combining several thin layers of pulp and squeezing through the press section. They are composed of a top liner, filler, and back liner. Bending grades of board will contain long fibers in the top and back liners sufficient to produce desired "bender."

Building of the board in layers provides the opportunity for filling the center of the sheet with low-cost waste papers. The filler of most newsboards is made of unsorted mixed waste papers which contain certain impurities. A solid board of this material is known as chip. Because of the unknown nature of the waste materials, cylinder

Courtesy of Akerlund and Rausing, Sweden

FIG. 5.4. PRINTED PAPERBOARD PACKAGE

boards are not usually used for direct food contact. Boards of this type most frequently used in packaging are bending newsboard, bending chip, solid Kraft-lined chipboard, bleached manila-lined newsboard, manila-lined newsboard or chipboard, and white vat-lined newsboard or chipboard. Vat-lined is the designation for a single-ply liner applied to the web by a cylinder machine, the tank of which is called a vat. The other special liners are also applied in this manner but may contain more than one vat application to improve color, strength, etc. Occasionally these sheets will be double lined; that is, a special liner is applied to the back of the sheet as well as to the top. For example, it might be white patent-coated, bleached manila-backed news.

Standard newsboards range in caliper from 14 through 35 points (0.014 to 0.035 in.). Available boards are: (1) White patent-coated newsboard, machine-finished. This is a high-quality printing stock, used largely for folding cartons where foil is on the interior of the carton, as the carton used for sample cigarettes. (2) Special lined bending newsboard, machine-finished, used for general foil-laminated folding carton work. (3) Special lined bending newsboard, clear vinyl-coated back. (4) Wash coated bending newsboard. Wash coated means that a lightweight coating has been applied by machine, to fill some of the voids in the surface of the board. This board has an especially good printing surface suitable for delicate tone work or fine detail.

The calipers for standard chipboards range from 14 through 35 points (0.014 to 0.035 in.). Available grades are: (1) Special lined bending chipboard, bleached manila back, used where a white carton interior is desired, but where strength and cleanliness of food board are not required. (2) Special lined bending chipboard, skim manila back. Skim refers to board similar to white vat-lined. This board is used for application similar to bleached manila except where an even poorer white lining is acceptable. (3) Special lined bending chipboard, Bogus Kraft-backed, used for applications similar to bending newsboard except where strength characteristics are required. It could be used for single-trip carriers, king-size detergent cartons, etc.

The white patent-coated and clay-coated boards have been developed to provide surfaces for high-fidelity printing. The surface smoothness or finish is determined by the formulation of the finish and amount of calendering. As weight (pressure) on the calendering rolls is increased, the board is compressed to high density, and a smoother surface is produced.

The National Paperboard Association recognizes four standard board finishes: No. 1, 2, 3, and 4. No. 1 is the most bulky (softest). No. 4 is the smoothest (densest or hardest) and has the best printing surface.

Corrugated Fiberboard

In 1871 Albert Jones was granted a patent for "a new and improved corrugated packaging paper." The earliest methods of making corrugated paper were shrouded in secrecy. It appears that the original concept was copied from the fluted rolls used in laundries for making ruffled lace collars. The earliest machines produced 20 ft of fluted paper per min about 2 ft wide. Machines are now capable of producing double-lined board 92 in. wide at speeds greater than 600 ft per min. In 1970 about 186 billion sq ft of corrugated were produced amounting to sales of over $3 billion, with volume and sales increasing annually. The industry has created new containers for a wide range of products. Yet, it has met the needs of longtime customers for inexpensive and dependable corrugated shippers.

Corrugated board is manufactured from three basic sheets—two liner boards and a central corrugated sheet or medium. These materials can be varied as to weight, type and number and/or height of the corrugations in the fluting medium. The three types of liner stock used are pure Kraft Fourdrinier fiberboard, test liner, and chip or straw papers usually for low-grade medium. Although the selection of a liner stock is extremely important, the properties of a corrugated board depend largely on the type, number, and position of the corrugations.

Made from straw paper, semichemical or Kraft paper, the medium or central corrugated sheet can be formed into four different flutes, i.e., A, B, C, and E. Most flutes used consists of A, B or the intermediate C. Flute selection is obtained by understanding the compressive strength required in the corrugated shipper. This takes into account the amount of weight on each container during shipment as well as the weight of the filled container. Flute A, which has the fewest corrugations per unit length, is an excellent cushioning material, has greater compressive resistance, but is not as easily bent as the others. Flute B, which has the most corrugations per unit length, can support greater weight than flute A when force is applied at right angles to the facings. Flute C is a compromise between A and B flute properties and so effective that it is widely used today.

The additional types of corrugated containers available include double-wall and triple-wall containers which contain two and three layers of corrugated medium respectively. A wide range of strength characteristics can be built into a triple-wall container since it is constructed from three corrugated medium sheets and four liners. The manufacture of corrugated board has not basically changed since 1890. Essentially the board is made by running the sheet fluted by exposure to steam through corrugated rollers and applying a facing sheet to each side of the flutes sheet by adhesive methods. The corrugated board is then usually printed by rubber plate or flexographic methods.

TABLE 5.2
CORRUGATED BOARD CHARACTERISTICS

Type	No. Flutes Per Ft Of Length	Flute Height (In.)
A	34–38, avg 36	3/16
B	50–54, avg 52	1/8
C	40–46, avg 44	5/32
E	85–95, avg 90	0.0050

BIBLIOGRAPHY

ANON. 1973A. Board—a miracle material for packaging. Emballages Dig. *16*, No. 154, 148, 153, 155-156. (French)

ANON. 1973B. Label paper. Mod. Converter, p.28.

ANON. 1973C. Improved techniques in the corrugated board industry. Packaging *44*, No. 523, 33-34,36.

ANON. 1974A. Papers, films, laminates. Emballages Dig. No. 169, 219-228. (French)

ANON. 1974B. Sweden beats board shortage. Packaging News Suppl. p.13.

BRISTON, J. 1973. Fibreboard competitive products—plastics. Converter *10*, No. 8, 20,22.

CLOTHIER, C. 1973. Packaging materials. Brit. Food J. *75*, No. 857, 178-179, 183.

DERRA, R. 1974. The importance of paper defined in the food law. Neue Verpack. *27*, No. 3, 326-328, 331. (German)

JOHNSON, F. D. 1973. Industrial waxed corrugated containers make major advances. Can. Packaging *26*, No. 8, 30-32.

MALTENFORT, G. G. 1973. Recent developments on structural aspects of corrugated board. Paperboard Packaging *58*, No. 8, 58A, 58C-58D.

MUTHUKRISHNAN, R. 1974. Physical tests for paper and board. Packaging India *6*, No. 2, 13-19.

PINNEY, G. 1974. What should you do when the board won't react properly to your starch? Paperboard Packaging *59*, No. 2, 26-27.

RAMAKER, T. J. 1974. Thermal resistance of corrugated fibreboard. Tappi *57*, No. 6, 69-72.

CHAPTER 6

Flexible Packaging—Paper

INTRODUCTION

Flexible packaging came into existence with the development of rapidly producing paper machines about a century ago. Mass production of paper started the industry known as "flexible packaging." Soon after the commercial installation of paper-making machines aluminum foil made its debut. In 1850 aluminum cost $545 per pound and was a curiosity valued above gold and silver. Barely 60 yr later the first experimental production of aluminum foil became a reality. It was not until shortly after World War I that aluminum foil was produced in commercial quantities. The first nonsynthetic transparent film, cellophane, appeared in 1924. Paper, aluminum foil, and cellophane presently constitute three of the four substrates used most by flexible packaging converters.

The coming of the plastic age gave converters three new films—pliofilm, saran and vinyls. Although each film has a definite place in converting, polyethylene brought synthetic plastic films into most converting plants. Discovered just before World War II polyethylene has made its mark in the last 20 yr. Along with paper, cellophane and aluminum foil, polyethylene is a member of the "big-four" webs. Polymer chemistry gave converters a wide range of additional synthetic films. Coupled with inks, coatings and machine innovations, it is theoretically possible to produce over 50 million different flexible packaging materials. The complexity of the flexible packaging industry is simplified by its characteristic of tailoring a material to a package. Converters must custom-laminate since no universal material exists.

A flexible packaging material is considered to be a material less than .010 in. thick. It is usually constructed of a composite or unsupported substrate based on paper, film and/or aluminum foil. The three basic materials used in the flexible packaging industry are discussed prior to a discussion of laminates and the fabrication of these constructions.

PAPER

Paper and packaging are a natural pair. Paper is the most important packaging material and packaging is the most important use for paper. Paper and paperboard together make up almost half of all packaging materials, while packaging accounts for about half of all paper production in the United States. Paper and paper products account for about one-third of the more than $20 billion spent on packages and containers each year. Perhaps the most important of the many papers available to the packaging industry is unbleached Kraft.

Kraft Paper

Strength is one of the outstanding characteristics of Kraft paper and accounts for its wide use throughout the industry. Over three million tons of unbleached Kraft is produced annually in the United States. Most of this helps to wrap or package everything from food to fertilizer. Four of the more important grades of Kraft paper are: (1) grocers' bag or sack paper, (2) shipping sack paper, (3) wrapping paper, and (4) gumming and asphalting paper. Although there are individual differences among these grades they have many characteristics in common. All are produced from the softer evergreen woods (such as fir and pine) by the sulfate process. All are considered coarse papers although their weights may vary quite widely. All of the Krafts are cheaper to produce than many other papers, because of modern high-speed production methods and the relatively low-cost raw material from which they are made. The Krafts are generally the most economical flexible packaging materials in use today.

In addition to these common properties each of the four major types has distinguishing characteristics. For example, grocer's bag or sack paper has the roughest finish, is quite porous, and has good tensile and tear strength. Shipping sack paper has a medium smooth machine finish to make it more receptive to the printing frequently applied to multiwall shipping sacks. Porosity is also important in shipping sacks. When such sacks are filled rapidly, particularly with powdery materials, it may be important that the air displaced by the incoming contents can escape through the paper rather than through the open mouth of the sack. Wrapping papers usually have a medium smooth machine finish, as do shipping sack papers, while gumming and asphalting papers are given even a smoother finish. The latter also generally have a higher tensile strength, and their thickness is more closely controlled.

Unbleached Kraft paper can be modified in many ways to produce varying characteristics for special applications. One of the more interesting methods results in extensible kraft, now finding wide use in the production of shipping sacks. The fibers of the paper are compressed by a heavy rubber blanket pressed against the web. Although the surface is not appreciably altered, the stretchability introduced into the Kraft increases its tensile strength and its ability to withstand impact. Coating, impregnating and laminating are three other ways that Kraft may be modified. Coatings include water-sensitive gums, for making gummed tapes and labels; polyethylene, to make the Kraft water-resistant; and various waxes and other hot melts. Wax may also be impregnated into the Kraft in the production of many types of bags and wrappers. Laminates include asphalt-Kraft combinations, polyethylene-Kraft, and aluminum foil-Kraft. Glass or nylon filaments may be introduced between the laminate layers for greater strength.

Bleached Papers

Unbleached Kraft retains much of the natural color of the wood pulp from which it was produced. For many applications, however, a white paper is essential. Bleached papers are generally made from a mixture of soft wood and hard wood, both by the sulfate and the sulfite processes. Earlier bleached papers were produced almost entirely from sulfite pulps, even though their final strength was lower, because the sulfate pulps were very difficult to bleach. Improvements in production methods, however, make it possible to produce the stronger sulfate papers in very high whiteness. Thus the unbleached Krafts have greater strength, while the bleached papers sacrifice some of their strength to achieve surface smoothness. This compromise is necessary because many of the bleached papers must be printed and modern high-speed printing methods require a very smooth surface.

As with the unbleached Krafts, the bleached papers are available with many different characteristics. They may be dense and highly polished, or they may be soft and almost fluffy. Their properties are varied to meet the two broad categories of application: (1) those that are largely functional and (2) those that are both functional and promotional. Functional applications include bags and wraps, such as butter wrap, freezer paper, butcher wrap, translucent wax paper and other food wraps. For these strength is the more important property; printability is seldom required. The whiteness of the paper denotes quality, cleanliness and purity. Those applications that are promotional as well as functional would include outside wraps and labels. The eye appeal demanded in highly competitive markets today requires that papers for these applications must have the maximum printability possible, while strength is somewhat less important. Multicolored designs are now almost universal for labels and wraps. The highest skill of the papermaker is applied to the production of papers that are suitable for such uses.

Glassines and Greaseproofs

In addition to the Krafts and the bleached papers many other types are available for packaging applications. Some of the more important types are: glassines, which may be plain, lacquered, waxed or laminated; greaseproof papers and waxed papers, which are distinguished by their coating rather than by the base paper; many grades of tissue papers; and the less paper-like packaging materials such as cellulose wadding.

The glassines are characterized by a glass-like smooth surface, high density, and (in most) transparency. Closely akin to the glassines are the greaseproof papers which are essentially uncalendered glassines. Together they are frequently referred to as G&G papers. Glassine and greaseproof papers are particularly suited to food packaging, which accounts for nearly 85% of production. They operate well on high-speed packaging equipment and take readily to printing. They are effective as odor and aroma barriers and can have wet strength if needed. Typical food applications include pouches for potato chips and dehydrated soups, primary packaging for cake and frosting

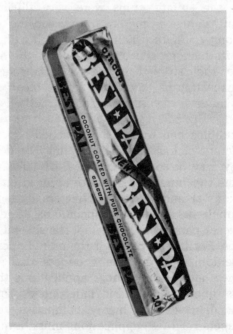

Courtesy of The Glassine and
Greaseproof Manufacturers
Association, Inc.
FIG. 6.1. GLASSINE FOR CANDY
Glassine is widely used for confectionary bar
packaging.

mixes, single-wall bags for bakery goods and ice cream bars, and duplex or multiwall bags for coffee, sugar and cookies.

Many of the waxed papers have their wax coating applied to a glassine base. In the wet waxed papers the coating remains on the surface of the base. In the dry waxed types the wax is actually absorbed into the base and the finished paper does not feel waxy. Glassine bases are wet waxed only, but sulfite and Kraft bases may be waxed either way. Waxed papers have many advantages as food wraps. Being tasteless, odorless, inert and nontoxic, they can come in direct contact with foods. They can be almost transparent, as in waxed glassine potato chip bags, or completely opaque. They have good protective qualities and heat-seal easily. They can be reclosed effectively when the contents have been partially removed. As with so many other packaging materials, the waxed papers have been improved in recent years. The 100% paraffin wax coatings have been modified by the addition of polyethylene, up to about 10% of the total coating weight. Other resins have also been used effectively, particularly ethylene vinyl acetate (EVA). Polyethylene permits much stronger heat seals and produces a more durable coating.

Tissues

The term tissue paper describes a thin semitransparent material used for gift wrapping. For years it was about the only paper available for this purpose. However, it has many drawbacks. It was never thick enough to effectively cover the gift, tore easily, was available only in white and a few solid colors, and did not take ink very well.

Modern tissue papers have changed all this. Tissue papers may be given strength in all wrapping directions and can be waxed for water-resistance. They are available in a wide range of colors and patterns. They may be impregnated with resins or chemicals that will inhibit tarnish of the wrapped item and can also be given spring-back after they have been crushed (for cushioning applications).

In the packaging industry tissues are generally considered as inner wraps. A good example is the waxed green tissue with which the florist lines flower boxes, and the unwaxed tissue around the blooms themselves. Hosiery and other fine textiles frequently are protected by an inner wrap of tissue. A coated hosiery tissue has been developed that can accept very fine printing and has a superior rich feel. Another specialized tissue is that used for wrapping fruit, particularly apples, oranges, and pears. Much of this tissue is impregnated with preservatives or with edible oils to prolong shelf-life and to increase product attractiveness.

In addition to its function as an inner wrap tissue paper is also quite effective as a separator. Panes of glass, sheets of metal, and layers of rubber or plastics are frequently separated by one or more sheets of tissue to prevent scuffing and sticking. Rolls of plastics are often interrolled with tissue, while waxed tissue may be placed between lacquered or printed surfaces to prevent damage.

One specialized application for tissue is in wrapping sterling silverware and fine jewelry. Ordinary tissue would not be satisfactory because its relatively high sulfur content would cause the silver to tarnish. A low-sulfur tissue is available, however, which may also have added chemical inhibitors which will further reduce tarnishing.

Other Papers

An even more effective cushioning material than tissue is creped cellulose wadding which has a soft, pliable, nonabrasive surface. It has three primary functions in packaging. The first is surface protection. The finest surfaces can be protected from scratching and etching by the wadding's ability to absorb microscopic dust particles into its open pores. Second, wadding is an excellent cushioning medium. Its light weight and resiliency help it to absorb impact in protecting delicate products from shock. The third function is as dunnage, to prevent item movement and damage during shipment.

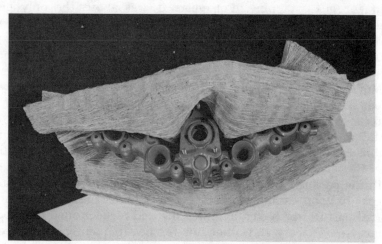

Courtesy of Kimberly-Clark Corp.
FIG. 6.2. CELLULOSE CUSHIONING MATERIAL

Paper Finishes

After paper is dried it is comparatively rough. Most paper machines are provided with smoothing rolls or calenders (highly polished steel rolls placed on top of each other) through which the paper passes. These rolls together constitute what is known as a calender stack. The surface produced by this calendering operation is known as machine finished (MF). The calendering is done as the paper leaves the dry end of the paper machine so that the paper is manufactured and calendered in the same operation. Most MF paper is of about the same smoothness on either side. However, sometimes the wire side will be a little smoother and sometimes the felt side.

Machine-glazed paper (MG) has a high finish on one side (the wire side) produced by a single huge dryer roll around which the paper passes, instead of several smaller ones generally used. This dryer has a very high polish and the drying must be controlled closely in order to get maximum smoothness. While machine-glazed papers may be further calendered, in most cases this is not necessary because the machine glazing operation gives high smoothness on one side. The paper machines making MG papers are often referred to as Yankee machines. The characteristic feature of this machine is the large dryer which is from 9 to 15 ft in diameter. The machine may have either a Fourdrinier or cylinder wet end and may have auxiliary dryers of the usual type.

Supercalendered paper (SC) has the highest finish or smoothness possible on paper. Unlike machine finishing, supercalendering is a separate operation. The calender stack for supercalendering is

Courtesy of Reynolds Metals Co.

FIG. 6.3. FLEXIBLE PAPER LAMINATE

This convenience package for cat food utilizes a flexible paper for its outer printed ply.

constructed on the same general principle as that used for machine finishing. It differs in that the rolls (usually an uneven number) are alternate steel and cotton. The bottom roll is always steel, the next roll cotton, the next roll steel, etc. The cotton rolls act somewhat as a cushion and have a certain amount of resiliency which would not be obtained with steel rolls alone. The best examples of super-calendered papers are glassines and coated litho papers.

English-finished paper (EF) has a smoothness between machine finished and supercalendered. It is generally low in gloss, which is accomplished by careful selection of finish and maximum machine calendering.

Water-finished paper (WF) or board has a high finish produced as it passes through the calender stack. Either one or both sides is moistened with a fine spray of water or by means of water boxes attached to one or more calender rolls. The water finish is not as uniform as a supercalendered and is usually confined to Kraft paper and paperboard.

When paper is processed so as to have a different finish on each side, it is referred to as duplex finished.

Coated Papers

Papers may be coated on one or both sides. The coating consists of pigments to which proper binders have been added to make the coating adhere to the paper surface. Clay or kaolin, a fine white clay used as a filler, is used more than any other pigment. Other materials are also used: titanium dioxide, zinc sulphide, aluminum hydrate, barium sulphate, calcium carbonate, satin white, etc. Among the binders used are casein, starch, latex and urea resins. The highest grades of coated paper are made in separate operations after the paper is manufactured on the paper machine. The paper can be coated by means of brush, knife, roll and other types, which apply the aqueous coating mixture uniformly to the paper. As the paper is comparatively rough after being coated, it is generally super-calendered. Great strides have been made in machine-coated papers during the last 15 or 20 yr. Today a considerable tonnage of paper is coated on the paper machine during the papermaking operation.

BIBLIOGRAPHY

ANON. 1973A. What is new in the region of paper packaging? Emballering. *4*, No. 8-9, 21, 23. (Norwegian)

ANON. 1973B. Paper waste becomes money. Emballering. *4*, No. 11, 19. (Norwegian)

ANON. 1973C. Finalizes 10 ppm limited for pcb level in paper for food packaging. Food Drug Packaging No. 3, 3,8.

ANON. 1973D. Glassine matches the challengers. Mod. Packaging *46*, No. 10, 29–31.

ANON. 1973E. Paper, kraft, wrapping. General Services Admin., Washington, pp.1–14.

ANON. 1973F. Paper-to-board lamination system is compact, fully automatic. Paperboard Packaging *58*, No. 12, 53.

ANON. 1974A. Packaging papers and boards. Emballages *44*, No. 314, 60–64. (French)

ANON. 1974B. Papers, films, laminates, Emballages Dig. No. 169, 219–228.

ANON. 1974C. Paper most popular. Paper *182*, No. 1, 25.

ANON. 1974D. Carton board from Finland: Quality and stiffness. Packaging Rev. No. 11, 66, 69, 71, 73.

ANON. 1975A. Paper-like films, synthetic papers and non-wovens (part I). Verpack. Rdsch. No. 3, 281–282, 284. (German)

ANON. 1975B. Stretching kraft paper for greater dimensions in converting. Paper, Film, Foil Converter No. 4, 42–43.

ANON. 1975C. Paper-like films, synthetic papers and non-wovens (part II). Verpack. Rdsch. No. 4, 402, 404–406, 408. (German)

CLOTHIER, C. 1974. Packaging materials. Brit. Food J. No. 857, 178–179.

DERRA, R. The importance of paper defined in the food law. Neue Verpack. *27*, No. 3, 326–328, 331. (German)

GHOSH, K. G. 1973. Development and application of fungistatic wrappers in food preservation. Part I. Wrappers obtained by impregnation method. J. Food Sci. Technol. No. 3, 105–110.

GOOCH, J. U., and PAINE, F. A. 1974. Developments in paper and board packaging. Paper Review of the Year, pp.111–112,115,118.

HEICHEL, G. H., HANKIN, L., and BOTSFORD, R. A. 1974. Lead in paper: A potential source of food contamination. J. Milk Food Technol. No. 10, 499-503.

NESHRAMKAN, R. H., and JAWARI, N. B. 1973. Role of kraft paper in packaging. Indian Pulp Paper *27*, No. 11, 18–19.

MUTHUKRISHNAN, R. 1974. Physical tests for paper and board. Packaging India *6*, No. 2, 13–19.

NORMENAUSSCHUSS, D. 1973. DIN 53103: Testing of paper and board determination of moisture content. Pira Translation *1218*.

PARDUCCI, M. 1975. How things stand with plastics paper. Imballagio No. 230, 29–40. (Italian)

PLACZEK, D. L., WITTER, K. T., and SCHINAS, E. 1973. Properties of plastics-coated papers and their evaluation as regards packaging technology. Verpackungs-Folien,-Papiere, pp.10–14, 16–20. (German)

ULLMANN, G. 1974. On the production of coated paper as a packaging material. Verpack. Rdsch. No. 11, 1078,1080,1083–1084.

VEALE, J. 1973. Waste paper as a raw material for paperboard manufacture. Pack. Print. Prog. *1*, No. 3, 2.

CHAPTER 7

Flexible Packaging—Films

A film may be defined as a flexible plastic (synthetic or natural) having a thickness of .010 in. or less. Although solid in its finished form, at some stage of manufacture it is capable of being formed into shape through the application of heat and pressure or by chemical reaction. A description of the most commonly used plastic films in the flexible packaging industry follows.

Cellophane

In 1908 the Swiss chemist, Jacques Brandenberger, tried spraying a tablecloth with a viscose solution in order to give it a smooth surface. He found that not only was the surface smooth but it also could be peeled off like skin. Three years later Brandenberger designed a machine to produce a material he called Cellophane. He devised the name from the first syllable of "cellulose" and the last syllable of the French word "diaphane" meaning transparent. Production in the United States began in 1924 in DuPont's Buffalo plant. Avisco started operations in 1930 and Olin constructed their first plant in 1951. In 1975 these three suppliers marketed about 270 million lb of cellophane, a large percentage of which is used by the food industry.

The outstanding commercial success of cellophane has been due to its adaptability. Over 100 varieties of cellophane are sold for various applications. They are constructed from the same base film by using different coatings. The first cellophane (PT) was moisture sensitive and dimensionally unstable. By adding a nitrocellulose coating the sheet became flexible and maintained a moisture vapor barrier. When an oxygen barrier was needed polyvinylidene chloride (PVDC) coating was applied to the base film. All cellophane suppliers offer a range of films from direct overwraps, converter use, carton or tray overwraps and bag applications. The nomenclature used for cellophanes differs for each supplier. Most cellophanes range between 0.8 to 1.9 mils in thickness. Instead of gauge designations the yield precedes the code describing the specific properties. A 195 MS cellophane has a yield of 19,500 sq. in. per lb and is moisture-proof and heat sealable. A 220 RS cellophane has a yield of 22,000 sq. in. per lb, is saran coated and is an Avisco film. If it were 220 K DuPont would be the supplier and for 220 V it would be an Olin film. The greater the yield, the thinner is the film. A 195 cellophane is thicker than a 220 grade cellophane.

Cellophane Manufacture.—Cellophane is made by a slot-die extrusion process. No resin is involved as in a thermoplastic film extrusion. Cellophane is made from wood pulp that is sent from the pulp mill in sheets to the cellophane film plant. The sheets are immersed in a 20% sodium hydroxide solution at 68° F for 30 min. After the excess liquid is pressed out, the alkali cellulose is sent to a shredder where it is made into small pieces and aged for 24 hr. The next step involves transferring the pieces to revolving drums. Addition of carbon disulfide follows. This mixture is then agitated for 2 hr until no residual vapor results. Transferring to a slurry tank follows and water is added. The sodium hydroxide solution is squeezed out in the alkali cellulose preparation. The net result is a viscose solution which is put through a mill to break up any small pieces, filtered and aged for about 24 hr. During the aging period degradation occurs and the viscose solution takes form.

Extrusion follows into a 10% solution of sulfuric acid buffered with sodium sulfate at 100° F. The die opening is about 50 in. long and is fully adjustable. By extruding into an acid bath the dissolved cellulose is made insoluble and regenerated. The web then passes into a hot water wash, alkali wash, cool water wash and then a bleaching and washing bath. The last bath constitutes a plasticizer and softener immersion in which glycerin or ethylene glycol and humectants may be added. The web is then wound up into rolls. Uncoated film is treated with a silica size to prevent blocking. For coated film a resin such as urea-formaldehyde is added to anchor the coating to the base web.

Uncoated cellophane is flexible, strong, transparent and greaseproof. It is also very hygroscopic and highly permeable to water vapor. When both sides of the film are coated with a moisture resistant lacquer (for example, nitrocellulose), greatly reduced permeability to water vapor and other gases results. There is still some slight permeation and for high-barrier packages, nitrocellulose-coated cellophane is not a satisfactory film. Nitrocellulose coatings are fairly economical and applied at rapid speeds. When the web is creased or roughly handled, however, the coating tends to fracture. For superior product protection, polyvinylidene chloride or PVDC-coated cellophane is commonly used. Often referred to as polymer-coated cellophane, the material has greatly superior grease and oil resistance. The coatings are not brittle and provide an excellent gas and moisture barrier. Applied in extremely thin laydowns (.05 mil) they offer good heat-sealing characteristics.

Courtesy of E. I. DuPont De Nemours and Co.

FIG. 7.1. WRINKLING PROBLEMS ARE ELIMINATED

T-1 cellophane overwrap offers excellent dimensional stability and means that it won't shrink and crush the pack under severe humidity and temperature conditions. T-1 cellophane offers excellent moisture protection and it greatly reduces the possibility of packs sticking together in shipment, warehouse storage and vending machines. Furthermore, it can be heat-sealed on standard machines without the use of solvents, thereby offering significant manufacturing savings.

PVDC is applied by a solvent system domestically. However, Europeans tend to favor the use of a PVDC emulsion.

Vinyl-coated cellophane was introduced by one U. S. supplier in 1963. Applied via a lacquer system, vinyl-coated film offers properties and cost intermediate to that of nitrocellulose-coated and PVDC-coated cellophane. In order to hold supermarket packages of bony chicken and meats, polyethylene (PE)-coated cellophane is also available. This material has an extremely thin coating of PE on the base web and is not to be confused with Polycel which is made by plastic converters. Useful in selected applications, PE-coated cellophane is rapidly gaining increased usage. Varieties of cellophane include improved release types, one-side-coated materials, different thicknesses and many other films.

Economics.—Cellophane is supplied in rolls and sheets. Rolls are priced by the pound and slit to order in various widths. The standard core size is 3 in. inside diameter. Widths range between 5/16 and 56 in. The three U. S. suppliers also offer other widths upon request as well as a variety of outside diameters. Sheets are priced by area in units of 1000 sq. in. They are cut to the desired specification. Stock size sheets are approximately 40 in. wide and 40 to 50 in. in length.

When cellophane was first introduced it sold at $2.65 per lb, a price which limited its market to the wrapping of luxury goods. Increasing markets coupled with improvements in manufacturing techniques led to dramatic price reductions. Between 1924 and 1939 there were 21 price reductions. In 1939 the average selling price for cellophane was $0.40 per lb. The rise in labor costs and raw material increases have led to a price increase over the 1939 level. By continuing improvements in technology this price increase has been held to about 170% above the 1939 low. Prices for most cellophanes now range between $1.08 and $1.13 per lb. Moisture-proof grades are the most economical with polymer-coated varieties being the most expensive. One basic problem with cellophane costs revolves around the number of different gauges of film produced. When the 250,000 sq in. yield polymer-coated material was introduced a thin economical film became available. Unfortunately this material yields fewer pounds per casting machine hour and thus is expensive to produce.

Cellulose Acetate

The origin of cellulose acetate film is related to the development of the first plastic, celluloid. In an effort to eliminate the flammability of celluloid (nitrocellulose), cellulose acetate was discovered. From Schuzenberger's experiments in 1865 to 1869 until World War I development was slow. Spurred on by fires in motion picture theaters (caused by the use of nitrocellulose as film) and the advent of World War I, the first commercial U. S. plant for cellulose acetate was built in 1918.

Overall usage of acetate film in packaging application has been steady during 1960-1970 at 5 million lb annually, decreasing to 2 million lb in 1975. The confectionery industry uses cellulose acetate fairly extensively, about 1 million lb or 20% of all cellulose acetate production in several different package concepts.

Manufacture of Cellulose Acetate.—Cellulose acetate, similar to cellophane, is a natural-type plastic material. While cellophane is made from wood pulp, cellulose acetate's raw material is waste cotton fibers known as cotton linters. The bales of linters are opened and the fibers are acetylated, using glacial acetic acid and acetic anhydride. Sulfuric acid in small amounts is used as a catalyst. The fibers dissolve slowly in these reagents and form a solution of cellulose triacetate. Water is then added to hydrolyze the triacetate and stop the reaction. Boiling increases the hydrolysis reaction and continues until a product forms with 2 1/2 acetate groups per glucose residue. The flakes of precipitated acetate are then washed and dried.

The resin formed by the above chemical reactions is made into film and sheet by two processes—casting and extrusion. In the casting technique the resin is dissolved in acetone with a suitable plasticizer such as dimethyl phthalate. Lower plasticizer contents can be used in resins that are solution cast as opposed to extrusion formed. The range of plasticizer content is 10 to 20% of the total resin weight. The dissolved resin-platicizer mixture is then cast onto a polished stainless steel band in a conventional casting operation. When all the solvent has evaporated the film is stripped from the rolls, dried and rewound for final usage.

Extruded film is produced by a conventional flat-die extrusion; however, relatively high plasticizer contents are necessary in order to obtain satisfactory melt flow properties. A small amount of solvent can be added and the solution is heated to about 230°F. Cellulose acetate can also be produced using an annular die process. The cast film is superior to the extruded variety in transparency, uniformity and surface smoothness. Extruded films generally exhibit die lines which can never be totally eliminated during processing.

For thicker film production (in the sheet range) cellulose acetate can be extruded in a finished shape. Generally 10 to 20 mil material can be extruded into various shapes from 2 to 8 in. long and up to 10 in. in circumference. This method allows for the resin to be formed into an end-product shape without any additional conversion.

Cellulose acetate resin can also be compression-molded, injection-molded, drawn and thermoformed. Prior to processing cellulose acetate usually must be dried to avoid bubbles caused by moisture absorption. This holds true for producing cast or extruded film as well as shaped extrusion products.

The film is not sensitive to moisture softening, is greaseproof, transparent and dimensionally stable, and ages satisfactorily. Cellulose acetate film is permeable to gas and moisture vapor and is currently approved by the FDA for direct contact with foods of all types. A wide variety of types are available in differing transparent grades for bags, while a low plasticizer content type is used for the more rigid window applications.

Semirigid Containers.—Cellulose acetate is also used in the form of a transparent semirigid container. It offers an aesthetically pleasing appearance and is easily fabricated. Cellulose acetate sheet in the range of 5 to 10 mils is commonly used. Equipment is now available that produces a fully finished acetate container directly from the cut sheet. The sheet is first slit to the desired size of flat sheet. It is then usually silk-screened, cut to container size, and rolled over a steel die to be sealed. A beaded edge is then used to improve the strength of the final container. Tops or bottoms may be transparent or made of paperboard, foil or other materials.

There are several distinct advantages for using acetate as a semirigid food package. Acetate does not crack or craze and the container may be custom-built around the product. Calendered, unplasticized vinyl (PVC) has made inroads into the semirigid plastic container industry.

As a thermoformed plastic sheet, cellulose acetate may be used in the form of a blister package. The range of sheet gauge used is between 0.003 to 0.02 in., either vacuum or pressure formed. The sheet is heated, drawn and stretched to the contour of either a male or female mold. Blister packages afford product protection and, if combined with a paperboard card, reduce pilferage in a supermarket. However, cellulose acetate is a poor moisture barrier, making it questionable for most food and drug products. Cellulose acetate is also a relatively high-cost material. In thermoforming cellulose acetate careful temperature control must be exercised. Attempts to form at low temperature lead to "milkiness" in the forming; high temperatures result in blistering and other defects. Cellulose acetate butyrate, a material similar to cellulose acetate, is often used as an alternative, but is significantly more expensive.

Laminations.—In composite flexible laminations requiring high glass and thermal stability, acetate is often laminated to other flexible substrates. A representative commercial laminate is (from the exterior in) acetate/foil/PE. The outer acetate film may be reverse printed, thus offering glass and scuff-resistance. Barrier characteristics are provided by the aluminum foil, while the polyethylene serves as the heat-sealing medium. A fairly recent trend has been toward the use of high-gloss coatings which offer superior gloss. This would tend to eliminate the necessity for acetate film. However, for increased body and rigidity and for protection of the barrier properties of the aluminum foil an outer film would be necessary.

Sealing Cellulose Acetate.—Cellulose acetate may be sealed by either heat or solvent methods. In heat-sealing either jaw-type or thermal impulse techniques can be employed. If jaw-type sealers are used cellulose acetate heat seals at temperatures between 375° and 450° F. A Teflon (DuPont registered trademark) coated bar is necessary to prevent the film from sticking to the bar. Teflon should also be used for a thermal impulse sealer. An impulse of electrical current is applied to a heat strip and sealing is effected rapidly and efficiently. Cellulose acetate is sealed best using thermal impulse methods.

Price of Cellulose Acetate.—Cellulose acetate in .001 in. gauge costs about $1.25 per lb or 5.7¢ per thousand sq in. For fabricated sheet packages, the price per pound generally remains constant. The cost per thousand sq in. is proportional to the gauge. In window cartons a section of the paperboard carton is die cut away and flexible cellulose acetate is used to cover the opening. The entire operation is fully automated. The surcharge for a window carton as contrasted to a plain carton runs about $1.55 per 1,000 cartons. In thermoformed or blister packaging the costs include the mold to form the blister, a heat-seal die and, if a paperboard backing card is used, the cost of the paperboard. Mold costs vary considerably depending on their complexity. The larger the number of cavities, the more expensive is the mold. A one-cavity mold may cost as much as $5000. A heat-seal die costs about $1,000.

Polyethylene

Two researchers at Imperial Chemical Industries in England running some experiments on ethylene found a waxy solid coating on the wall of the pressurized vessel. From this event the first low-density polyethylene became a reality. After World War II refinements appeared and medium- and high-density polyethylene were introduced.

Until the petrochemical crisis, close to 1 billion lb of polyethylene was sold annually with considerable growth projected. Basic resin technology of polyethylene is well established, production methods are well developed and production equipment is available. Polyethylene consumption rose from 175 million lb in 1958 to 900 million lb in 1969, an increase of over 500% in 11 yr. In 1975, the

TABLE 7.1
POLYETHYLENE PROPERTY CHART

Type of PE	Moisture Vapor Transmission[1]	Gas Transmission[2] O_2	CO_2	Tensile[3] Strength	Heat Sealing Range
Low-density	1.4	500	1350	17.0	250–350°
Medium-density	0.6	225	500	25	250–350°
High-density	0.3	125	350	40	275–350°

[1] gm/100 sq in./24 hr/76 cm Hg at 100°F, 90% RH for 1 mil film
[2] cc/100 sq in./24 hr at 72°F for 1 mil film
[3]
[4] 100 psi
Moisture effects gas transmission rates of cellophane and may vary when wet.

volume was 1150 million lb. Most applications involve low-density film, with a considerable market also existing for high-density resin for molding purposes. A new market for special second generation blends called ionomers has recently appeared.

About 40% of the world polyethylene production is converted into films of .010 in. or lower gauge. About 75% of all polyethylene film manufactured is used for packaging. Polyethylene film is available to the packaging industry as: low density, medium density, high density, and shrink film.

Low-density Polyethylene.—Polyethylene is a crystalline polymer with density varying between 0.915 and 0.970 gm per ml. Its optical properties, coefficient of friction, resistance to blocking and printability are dependent on the extrusion conditions, modifications and subsequent treatment of the polymer. As polyethylene density increases, better resistance to oils and fats occur. Overall resistance to fat and oils is good, however, high-density film offers somewhat better protection. In addition, the permeability of polyethylene to

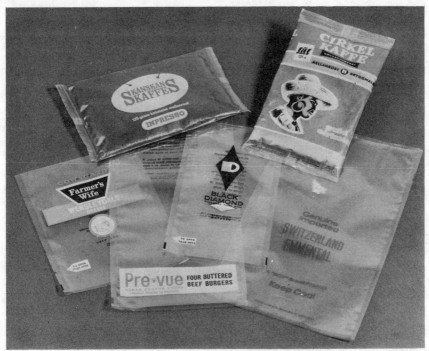

Courtesy of Imperial Chemical Industries, Ltd.
Plastics Division

FIG. 7.2. POLYETHYLENE LAMINATE
This all film laminate uses LDPE as the inner heat-sealing ply.

water vapor and gas decreases as density increases. All polyethylene films are excellent water barriers, but fairly poor barriers to nitrogen, oxygen and carbon dioxide. Polyethylene is inert to most chemicals. However, animal, vegetable and mineral oils are absorbed by polyethylene and may cause it to swell and discolor.

Properties.—Three principal characteristics are used to classify a polyethylene resin: density, molecular weight, and molecular weight distribution.

Density is a measure of crystallinity and directly influences the properties of the polymer in the solid state. As crystallinity increases so also does density, but all properties dependent on crystallinity decrease. Thus, water vapor transmission rate (WVTR), gas transmission and grease penetration decrease with an increase in polyethylene density. The three general density ranges for polyethylene are: low—0.910 to 0.925 gm per ml; medium—0.926 to 0.940 gm per ml; and high—0.941 to 0.965 gm per ml. To the processor, as the density of a polyethylene resin increases, melt viscosity, draw-down and resistance to blocking increase.

Molecular weight is measured by melt flow index and melt viscosity. Melt flow index is defined as the flow through a standard die under standard conditions of temperature and pressure. If the melt index of a resin is low, its melt viscosity is high and vice versa. Resins with a higher melt flow index flow more easily in the hot state than those with a lower melt flow index. A resin with a melt flow index of 6 will not flow as readily under a heat sealer as one with a melt flow index of 11. The heat seals of a polyethylene resin (MI = 11) are extremely strong. In processing polyethylene melt flow index is important because all the operations involve melting the resin and then converting the melt to a specific item (i.e., film, coating or molded piece). It is important to note that the flow characteristics of molten PE are also greatly affected by machine conditions. While melt flow index is a valuable parameter it is not all-inclusive.

Molecular weight distribution is a measure of the ratio of large, small and medium molecular chains in the resin. If the resin is composed of chains of similar length its molecular weight distribution is narrow. A resin composed of a great variety of lengths is referred to as one with a wide molecular weight distribution. Polyethylene resins with a narrow MW range are superior mainly in barrier properties, processability, and stress-cracking to those with wide ranges. Impact strength is also dependent on molecular weight distribution. Narrow MW distribution yields the toughest film especially when coupled with a low melt flow index.

Since polyethylene is produced by one of three extrusion techniques, neck-in and draw-down are also important characteristics. Neck-in is caused by surface tension and melt elasticity effects operating in the immediate vicinity of the die exit. Low neck-in is desirable because there is less film waste in the edge trim. Draw-down is defined as the coating speed (ft per min) beyond which the melt tears in the vicinity of the die. Both neck-in and draw-down are dependent to a great degree on swelling ratio. The latter is a measure of melt elasticity. If the resin has a high swelling ratio, neck-in is reduced. But with a high swelling ratio draw-down is decreased and a thin film becomes difficult to produce. This is but one example of the complexities inherent in resin selection. In selecting a suitable resin for polyethylene production, all parameters are interrelated and must be carefully considered in light of results desired and equipment conditions.

Polyethylene Production.—The manufacture of polyethylene film depends upon the characteristics of the resin and the properties desired. Polyethylene film is made by three methods: (1) extrusion blown tubing, (2) extrusion onto a chill roll, and (3) extrusion into a water bath. The first method produces extruded blown film; the latter two produce cast films. In producing polyethylene film a conventional screw extrusion process is used. A circular die is used for blown film while a slot-die process is generally used for chill-roll and water-quenched film.

Blown film is usually extruded upward from a circular die and air pressure expands the tube while it is still hot. As the polyethylene moves upward it cools and near the top of the tube is flattened, guided over rollers and brought down to the winding stand. By nipping the film after it has been cooled and applying an excess air pressure to the interior the tubular blown film can be formed to a diameter greater than that of the die. This ratio is known as blow ratio. The die diameter and rate of blow can also be controlled. By rotation or oscillation of the extrusion die thickness variations and high spots are reduced.

Slot-die (cast) produced film is generally cast onto water-cooled highly polished metal drums. It is also possible to quench the melt by passing it through water. As the melt comes out of the die it is drawn into the nip between the pressure roll and water-cooled chill roll (or quench tank). The film is either obtained unsupported or forced onto a substrate for use as a coating.

In blown film there is a strong relationship between the impact strength and the tear strength in the machine direction. The film impact strength is governed by the balance of tensile strength and

TABLE 7.2
POLYETHYLENE USAGE CHART

Commodity	Packaging Characteristics Needed	Type PE	PE Packaging Machines[1]	Vol PE
Bakery products (bread and cake)	Moderate moisture barrier; grease-proofness for cakes, etc.	LDPE film (bags) —1¼ mil for bread	AMF Mark 50 Pneumatic Scale Corp. FMC—"Form and Fill"	85 million lb
Biscuits and crackers	Excellent moisture barrier; grease resistance; protection from physical damage	Cracker (slugs) in PE bags-LDPE; PE too soft for use in Hi-speed cracker packaging	AMF Mark 50	Low
Cereals	Sweetened require moisture protection; fragile—protection needed; possible protection from rancidity	Paper/LDPE pouch; multi-packing with PE film	Bartelt Delamere & Williams Bundling machines	4 million lb
Coffee and condiments	Coffee—good protection from moisture and O_2; condiments—moisture protection, odor and flavor retention	PE snap-on caps for cans; polyester/foil/ MDPE pouches; paper LDPE pouches	Bartelt Stokeswrap-vac Hesser SIG	Moderate
Confectionery	Excellent—poor moisture protection; odor protection	PE bags (.001 -.0025 in.); multi-packs; Overwraps-LDPE; Surlyn "A"	Hudson-Sharp Campbell Transwrap Hayssen	20 million lb
Convenience foods	General guidelines apply; odor retention	LDPE laminates of paper and foil; film overwraps	Circle Pouch Maker Cloud Machine Bartelt	14 to 18 million lb

Dairy products	Grease and odor protection; moisture loss; O_2 protection	Blow molded PE bottles and PE coated board; PE bags; PE-foil laminates	Thimmonier Hayssen Hudson-Sharp	12 million lb
Dehydrated foods	O_2 protection and moisture; fragility	PE bags; PE laminates	Bartelt Delamere & Williams	15 million lb
Desserts and dessert mixes	Moisture protection; flavor retention	PE coated wax-laminated glassine; PE laminates	Bartelt Delamere & Williams	Moderate
Frozen foods	Good moisture barrier; Low temp durability; O_2 barrier	LDPE film (bags)—3 mil; LDPE laminates with film and foil	Hayssen Bartelt Package machinery	19 million lb
Meat	Fresh—O_2 barrier poor, good moisture barrier; cured—good O_2 barrier, good moisture barrier	LDPE film overwrap; laminates with film and foil	Std. Pkg. Models Mahaffey & Harder Hayssen	20 million lb
Poultry and eggs	Good moisture barrier; good O_2 barrier	LDPE film (bags) for whole birds; LDPE overwrap	Hobart Package machinery Cleveland-Detroit	Figure above includes poultry
Produce	Breathable film; transparency; nonfogging	LDPE film (bags) —perforated; PE netting	Conventional bagging machines and manual models	160 million lb
Snacks	Good O_2 barrier; good odor barrier; good fat barrier	PE laminates with PP, cello; PE bags	Woodman Wright Mira-Pak Others	2 million lb

[1] These machines are representative models—the list is *not* all-inclusive.

elongation in the machine and transverse directions. Since the molecules in a blown film tend to become aligned in the machine direction, tubular film is weaker at this point in impact strength than slot-die film. If a high extrusion temperature is used superior optical and mechanical properties result, but the tendency to block increases. If a high blow ratio is used an increase in both optical and mechanical properties also result. Bubble problems, however, often lead to breaks and production difficulties.

The cooling rate controls film crystallinity, thickness and output rate. As the rate of cooling increases, a more impact resistant and smaller crystal size film is produced. Minimal crystalline irregularities and size produce the most glossy film. Also the location of the line, height and bubble shape influence the film's properties. Slot-die film is not as influenced by machine direction orientation as blown film. However, impact strength is greater in the machine direction. By the use of a high melt temperature surface problems are minimized. Smaller crystals result froom a rapid cooling rate.

Blown and Cast Film.—Optical properties are dependent on the cooling rate during extrusion. Cast film has optical properties superior to blown film since the rate of cooling for cast film is greater during production. An optically acceptable blown film can be produced by careful control of melt temperature, blow ratio, output rate and mixing. Guidelines are the need for good screw design, high-temperature extrusion, bubble geometry control and high output rate. In cast film, transparent film can be made easily from high-density resins.

Production economics favor blown film which is cheaper than cast film because the die used is easier and more economical to produce than a slot-die. In making wide-width films a blown film is substantially cheaper than a slot-die film. Wide-width cast film also involves the use of very wide and expensive chill rolls.

Although the impact strength in the machine direction (MD) of a blown film is weaker than that of a cast film, overall mechanical properties are greater in blown films than in cast. During the extrusion process blown film becomes oriented, resulting in improved tensile strength. Shrink polyethylene films can be produced and (if nonshrink is desired) the film can be heat-set. Further, there is no edge trim loss in blown film as there is in cast.

Cast film can be produced at higher production speeds than blown film. Its gauge variation is less than blown film.

Additives.—Plasticizers are almost never added to polyethylene. To minimize the development of odor in polyethylene a trialky phenol may be used. In packaging machinery good slip is necessary in

Courtesy of E. I. DuPont De Nemours and Co.

FIG. 7.3. SKIN PACKAGE

Surlyn is used to skin package this hardware.

providing increased speeds. Slip additives are added and care must be provided to prevent bloom to the surface of the film. Antioxidants are added to prevent embrittlement from ultraviolet light rays. Recently there has been some development in adding materials to polyethylene film to provide easier visible light degradability. This is an attempt to improve polyethylene's position in the ecological system. To minimize dust attraction antistatic compounds are also added.

Polyethylene Modifications.—Ionomers are constructed of both covalent and ionic bonds. The introduction of Surlyn A by DuPont marked the development of this new family of resins. Based on low-density polyethylene, the ionic bonds in ionomers serve to increase overall bond strength and yield superior oil, grease and solvent resistance.

By copolymerizing low-density polyethylene with vinyl acetate ethylene-vinyl acetate (EVA) is obtained. EVA is more flexible than polyethylene but may tend to block due to high surface friction. Slip and antiblock additives partially prevent this tendency. The properties of EVA copolymers are related to the percentages used of polyethylene and vinyl acetate (VA). A 90 (PE) - 10 (VA) blend produces a close resemblance to conventional low density poly-ethylene. A 70 (PE) - 30 (VA) mixture takes on the physical characteristics of gum rubber.

Polyallomers are combinations of polyethylene and poly-propylene. They reportedly yield better low-temperature character-istics than either single constituent. Most of their properties fall between high-density polyethylene and polypropylene.

Interesting modifications of polyethylene film also include poly-hydroxyalkanes, based on hydrolized EVA materials, still in the research stage. Available commercially is irradiated PE which is produced by irradiation of low-density polyethylene with high-energy electrons. The result of this process is to produce a number of cross-linkages between the polymer chains, which serve to increase the strength and heat resistance, as well as altering the permeability to gases and moisture vapor.

Courtesy of Montedison, Italy

FIG. 7.4. FERTENE (MONTEDISON POLYETHYLENE) LAM-INATED PAPER SHOPPERS FOR BOUTIQUES

Polypropylene

In 1954 Professor Guilio Natta of the Polytechnic Institute in Milan succeeded in polymerizing propylene to high molecular weight solid polymers. Using sterospecific catalysts his discovery of Isotatic polypropylene (PP) aroused great interest in the packaging industry. It was not possible to regulate the growth of polypropylene chains in exactly the spatial altitude required. Initial production of cast polypropylene film occured in 1958 and the various oriented grades soon followed. With over 120 million lb of polypropylene film manufactured annually consumption appears to be rapidly growing. Capturing markets from the nonthermoplastic cellophane, and characterized by many different types, polypropylene is a versatile and economical film.

Properties of Polypropylene Films.—All polypropylene films are constructed from a highly crystalline polymer with a permeability about 1/4 to 1/2 that of polyethylene. They have good resistance to acids and alkalis and at room temperature are unaffected by most hydrocarbon solvents. At elevated temperatures solvents attack polypropylene. Stiffness, clarity and grease resistance are also greater than polyethylene. The impact strength of polypropylene is not as good as that of polyethylene, but it has superior tensile strength.

TABLE 7.3
FOOD PACKAGING FILM

| Polyethylene | Greases & Oil | Resistance To: | | |
		Acids	Water	Alkalies
Low-density	Fair	Excellent	Excellent	Excellent
Medium-density	Good	Excellent	Excellent	Excellent
High-density	Excellent	Excellent	Excellent	Excellent

Polypropylene films are available as: cast; oriented, heat-set; oriented, shrinkable; coated; and composite. Many variations exist within each category. Oriented heat-set films may be uniaxially or biaxially oriented. Coated films may contain a saran (PVDC) or heat-sealing coating.

Cast Polypropylene.—Cast polypropylene is made by a slot-die extrusion process using a die opening between .010 to .020 in. After passing an air gap of several inches the web is cast onto highly polished chill rolls. It is then passed through a nip roll onto various idlers and to the final rewind station of the extruder. Extrusion temperatures of 550° to 610° F are common.

Rapid cooling of the extrudate is essential. If the molten resin is not cooled quickly crystallites form which lead to the growth of spherulites. The net result would be opacity in very thin films (.006 to .012 in.). Quenching involves the use of chilled rolls to cool and freeze the molecular structure into the structural pattern it has assumed at a particular point of manufacture. Cast polypropylene film has outstanding clarity because the molecular chains of the film are frozen before they have the opportunity to realign themselves in crystalline configurations. Quenching does not permanently prevent the danger of crystallization, but it retards its rate of growth significantly at room temperature. The opacity of cast polypropylene film increases with increasing thickness because rapid chill-roll cooling becomes more difficult to perform during extrusion.

Cast polypropylene is characterized by good stiffness, grease and heat resistance, and provides an excellent moisture barrier. However, cast polypropylene is not an outstanding gas barrier and may become brittle at low temperatures. Its tensile strength is approximately double that of cast polyethylene film, but its s tear strength is about one-half that of polyethylene. Two types of cast polypropylene are available, a homopolymer and an ethylene-propylene copolymer. The differences relate to packaging machinability and heat sealing. The copolymer tends to be softer than the homopolymer. However, both are identical in optical properties.

Orientation.—Orientation involves the stretching of the polymeric chain of a plastic to generally improve the properties of a film by either thermal or mechanical methods. An orientation process can be performed either longitudinally or laterally. If performed in one direction only the orientation is known as uniaxial; when performed in both directions it is called biaxial. In biaxial orientation balance relates to the degree of stretching in both directions. A balanced film is stretched equally while an unbalanced film is stretched differentially in both directions.

Two methods are used to manufacture an oriented polypropylene film: a bubble process and extrusion followed by mechanical stretching.

The bubble process involves extruding a tube of polypropylene, rapidly quenching to 32° F, reheating to about 212° F, and expanding the tube under air pressure while it is being axially accelerated 2 to 10 times by the final nip rolls. Draw ratios are adjusted by the volume of air in the bubble and nip-roll speed. In the mechanical stretching technique thick cast film is first produced by slot-die extrusion. The film is then fed directly through a system of stretch rolls where longitudinal machine direction (MD) orientation is performed. Two stretch rolls, each at different speeds, are used. The rolls are heated and the film is softened by the first roll. It is oriented

Courtesy of Imperial Chemical Industries, Ltd.
FIG. 7.5. ORIENTED POLYPROPYLENE PACKAGE
ICI's Propafilm is used to package snack foods.

by the second roll which is driven at a faster speed than the first. After leaving the MD stretcher the film passes through pinch rolls directly into a tenter frame. In this operation lateral cross-direction (CD) stretching is performed. The tenter frame consists of a system of driven chains which usually carry film through a hot-air oven, cooling zone and rewind rolls. By means of a series of clips mounted on the chains of the tenter frame, opposite edges of incoming film are gripped for lateral stretching. Accurate temperature control is necessary during the entire operation. The film may first pass through a conditioning process in which the clips proceed parallel to each other. Next is the hot-air oven where the film is heated to orientation temperatures and where stretching is affected by the ever widening gap across which the film is drawn by the chains. The film then passes through a cooling zone, the clips are released and it is fed into the rewind station. If either step of this sequential system is omitted a uniaxially film results. Speeds up to 500 ft per min with .0005 in. polypropylene are possible using a tentering technique. Work has been done in developing a system which can effect both MD and CD draw in one operation.

The properties of oriented polypropylene produced by either orientation method are fairly similar. The bubble process appears to produce a better balanced film. However, the sequential operation produces a film with better tensile strength. Both processes produce acceptable films and one is not clearly superior to the other.

Heat Set-oriented Polypropylene Films.—For nonshrink oriented polypropylene films, the orientation process is followed by an additional heat treatment at a higher temperature while the film is held under restraint. The film loses its ability to shrink and remains stable although the degree of orientation and the optical and physical properties remain unaltered. Heat-set film forms because of the occurrence of crystallization and the locking in of the oriented molecular chains in the oriented polymer. Heat set-oriented polypropylene films are characterized by good low temperature durability, high stiffness and excellent WVTR.

One of the most controversial areas relative to heat set-oriented polypropylene film is the relative merit of balanced versus unbalanced film. In bags balanced film tends to tear more in one direction when tear is initiated; unbalanced film tears at an angle. On the other hand, the tensile strength of balanced film at below freezing conditions is superior to that of unbalanced film. It is important to note that in laminations this difference is minor. Both films perform satisfactorily under most conditions.

Uniaxially-oriented films tend to fibrillate when subjected to stresses at right angles to the orientation direction. This does not occur in either balanced or unbalanced film. In almost all their physical properties biaxially-oriented polypropylene films are similar to cellophane.

Shrinkable Polypropylene Films.—Shrinkable polypropylene film is not heat-set after it is oriented. It is used for shrink packaging in which it shrinks tightly around a package after the entire unit passes through a hot-air tunnel. Polypropylene is one of the least expensive of the shrink films. It has good WVTR, fair gas permeability and excellent moisture and grease resistance. The maximum amount of shrink possible is dependent on the degree of stretch performed during orientation. Polypropylene is able to shrink between 60 to 80% at hot air temperatures of 425°F. It can be heat-sealed at 350°F. Most shrink films are equally balanced as to orientation although several specific packaging applications may involve an unbalanced shrink film.

Crystallinity may be a problem with shrinkable polypropylene films. Unless the minimum practical heat-sealing temperatures are used crystallinity will effectively weaken the finished seal to a fraction of its original strength. An additional factor is that the shrink temperature for polypropylene film is fairly high and is in the range of the film's melting point. These high temperatures must be carefully controlled in order to prevent a cloudy and weakened film.

Coated Polypropylene Films.—In order to enhance further the heat-sealability, barrier and specialty requirements of polypropylene, coatings can be applied to its surface. Oriented polypropylene is usually coated with an aqueous system since it tends to be sensitive to solvent applications. A primer is also used to develop satisfactory

Courtesy of Imperial Chemical Industries, Ltd.
Plastics Division

FIG. 7.6. PVDC-COATED POLYPROPYLENE FILM
Used for sweet biscuit packages.

adhesion between the coating and film. Two varieties of coated, oriented polypropylene film are available: heat-seal coated and saran (PVDC)-coated. Coatings are applied by one of several methods, i.e., air knife, gravure roll, blade or metering roll coater. Heat-seal coatings on polypropylene are based on polyvinyl acetate in an aqueous suspension. This system yields excellent heat-seal properties but there is no barrier improvement. The use of a heat-seal coating on polypropylene film provides for a broader heat-sealing range and greater film versatility.

Saran-coated polypropylene, also known as barrier-coated, greatly improves the gas barrier properties of the base film. Its permeability to moisture is reduced by a factor of 2 while gas permeability is reduced from 1,800 to 6 cm^3 per m^2 per 24 hr at 1 atm pressure differences.

Composite Films.—By simultaneous extrusion of resins through complex dies, coextruded or composite polypropylene films are produced. Intermixing of the polymers is avoided through the maintenance of laminar flow throughout the molten phase. Both flat die and blown coextruded films are now being manufactured.

One of the first coextruded films available was a PE-PP-PE structure designed for breadwrap. Developed by Kordite Corporation (now Mobil) in 1963, the film has a total thickness of one mil consisting of .0004 to .00045 in. PE—.001 to .0002 in. PP—.0004 to .00045 in. PE. Medium-density polyethylene is used for both outer piles. This material is readily heat-sealed and has excellent low-temperature flexibility.

Other types of coextruded polypropylene films have recently appeared. Ethylene vinyl acetate (EVA) PP and EVA-PP-EVA are now being used for frozen food overwraps. Coextruded films are easier to make when one of the members is a polyolefin.

Additives.—Additives are now widely used in polypropylene resins. Antistatic agents such as polyethylene glycol may be used but surface bloom may become a problem. For protection against untraviolet radiation stabilizers can also be used. Slip additives are sometimes incorporated to provide ease in machinability.

Polystyrene

The discovery of styrene is generally attributed to E. Simon in 1839. Polystyrene was first synthesized in 1866. The capture of Far Eastern rubber plantations by the Japanese during World War II provided the incentive necessary for large-scale commercial development of the resin. The synthetic styrene rubber (Buna-S) was developed. Oriented polystyrene (PS) film introduced in 1948 achieved large-scale use by the mid-1950's. With the development of high-impact grades, acrylonitrile butadiene styrene (ABS) types and other varieties, polystyrene is one of the most useful packaging films. Over 15 million lb of polystyrene film is produced annually for a variety of applications.

Properties of Polystyrene Films.—Polystyrene film is available in unoriented, modified, oriented grades and as a cellular material. It is resistant to acids and alkalis and is not affected by the lower alcohols or glycols. Aromatic hydrocarbons and higher alcohols attack the film. However, the film exhibits extremely poor WVTR and gas transmission rates. Elongation and tear strength are also poor. On the positive side, the films have excellent clarity, sparkle and gloss. They are FDA approved and are widely used in food packaging.

Unoriented Polystyrene Film.—Unoriented polystyrene film is a brittle inflexible material. Until the introduction of orientation polystyrene could not be used as a free film and was restricted to use in laminations. Made by a conventional slot-die extrusion process, polystyrene must be modified prior to use in packaging (for example, by a tentering process).

Modified Polystyrene Film.—If polystyrene is produced with minor proportions of synthetic rubber added, a much tougher and more flexible film results. The rubber serves to prevent crack propogation and increases extensibility. However, there is some sacrifice of transparency and tensile strength. Rubber-modified polystyrene film is known as toughened polystyrene, impact, high-impact and super-high-impact, depending on the properties of rubber to styrene.

Using a slot-die extrusion technique carefully formulated polystyrene pellets are first thoroughly dried. Extruder screw design is essential in maintaining a high production rate. Constant depth, decreasing pitch screws are less useful, since they tend to limit production by trapping air. A 20 mil die opening is used to produce 0.5 to 2 mil thick film. Production rates may range between 400 to 800 lb per hr. It is also useful to carefully control the polymer mix and thickness during extrusion.

Since the addition of small amounts of rubber effects the transparency of polystyrene film, several techniques have been developed to increase gloss on the extruder. By using infrared heaters, either in front of the extruder die or behind the rewind rolls, heat glazing occurs. As the film is extruded it is carefully heated in order that minimal change occurs in the overall mechanical properties of the film. If highly polished rewind rolls are used with a narrow gap between them, the pressure on the freshly extruded film effects some degree of gloss. A combination infrared and pressure roll system will provide for the best gloss possible. Copolymers are also available with ABS (Acrylonitrile-butadiene-sytrene). The resultant film is always tougher than the homopolymer but with a sacrifice in transparency.

Oriented Polystrene Film.—In order to improve the inherent brittleness of pure polystyrene without the addition of synthetic rubber, the film may be biaxially oriented. While the optical properties of polystyrene depend on the rate of cooling during extrusion, the physical and mechanical properties are increased by orientation. Improvements become evident in elongation, impact strength and toughness, without any sacrifice in gloss or transparency. Three methods are used to produce biaxially-oriented polystyrene film: bubble, tentering and radial stretching.

In the bubble process a stream of molten polystyrene is forced through a rotary cooler by means of a gear pump. This reduces the polymer temperature from about $392°$ to $284°F$. Extrusion then follows via an annular die through a cooled guide ring and the extruded tube is expanded 6 to 8 times around an air bubble which induces transverse orientation. By using about .25 psi air pressure the entrapped air bubble acts as an air mandrel quite well. The film is stretched at the same time by take-off rolls which pull at a linear rate 6 to 8 times greater than the extrusion rate at the die. Longitudinal orientation is thus induced. There are two inherent advantages in producing biaxially-oriented polystyrene film by a bubble process. As the hot extruded tube expands and accelerates downward it is also cooling. When its ability to flow is reduced it is cooler, and with little control over various processing variables orientation is introduced. By using an internally cooled guide ring after the die the top skin of the extruded film is rapidly cooled to the required level and provides a more uniform film. The air ring also serves to keep the bubble centered. The bubble process is only applicable for gauges less than .003 in.

The tentering process involves the use of an extruded polystyrene sheet which has about 90 mils gauge and is 15 in. wide. Extruded through a slot die at $220°F$, it is passed around two chilled polished rolls and temperature is reduced to about $230°F$. It is then passed through a drafting section where longitudinal stretch is induced by two nip rolls. Film stretch of 300% is effected by keeping the speed of the second nip roll three times that of the first. After the film is cooled to about $220°F$ it is fed into a tenter frame maintained at $230°F$. A series of clips hold both edges of the film. The clips move in a diverging pattern and cause the film to be stretched in the transverse direction. The next steps are cooling, clip release and rewinding. By using this process, a ninefold reduction in final thickness of the extruded sheet is obtained and an equalized threefold stretching is possible. Complete balance is properly controlled by the tentering process. A wide variety of gauges of finished film can be produced by tentering since the ultimate film thickness depends on the initial caliper of the extruded sheet. While tentering is a complex process characterized by the need for careful control, it is also commonly used.

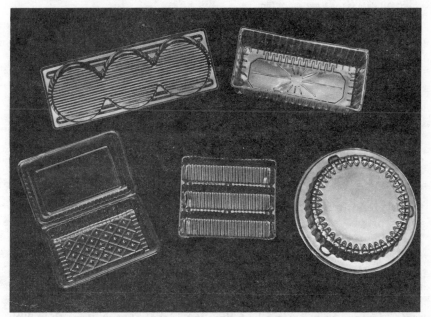

Courtesy of Rexene Polymers Co.

FIG. 7.7. BIAXIALLY-ORIENTED POLYSTYRENE SHEET

Radial stretching is the process used least for producing biaxially-oriented polystyrene film. A flat sheet is extruded as a diaphragm and is then stretched by several pairs of take-off rolls around the periphery. The process closely resembles a conventional tenter frame with the omission of any edge gripping mechanisms.

Heat Shrinkable, Oriented Polystyrene Films.—By orienting polystyrene film heat shrink characteristics are obtained. Shrink occurs when the film is heated to temperatures near the stretch temperature used. Polystyrene shrink film exhibits shrinkage at 225°F up to 40 to 60%. It is an extremely cheap shrink film characterized by high sparkle, clarity, permeability and gloss. However, it is brittle and

TABLE 7.4
PERMEABILITIES OF VARIOUS FILMS

Film	WVT[1]	Oxygen[2]
PVDC	1.5-5	8-26
Low-density polyethylene	18	4,000-13,000
Oriented polystyrene	>100	2,000-7,700
Polyamide	Extremely high	30-110

[1] gm/24 hr/1 sq m/100°F, 90% RH
[2] cc/mil/1 sq m/24 hr/1 atm/73°F, 0% RH
Note: All data is for 1 mil gauge film.

easily marred. It has been used successfully in packaging fresh produce, since the film allows the products to "breathe." The hard metallic sound of the film is an intangible asset to the consumer, who appears to associate the sound with a quality image.

Polystyrene Foam

Introduced by Dow Chemical Company in 1943, foamed plastics offer excellent properties to the package user. Although the basic chemistry involving foamed plastics originated in Germany, it was first refined and commercially used on a large scale in the United States. Over 1 billion lb of foamed plastics are used annually and polystyrene foam accounts for about 25% of total production. Two types of polystyrene foam are available: open cell and closed cell. The manufacturing processes used in their production differ considerably.

Properties of Polystyrene Foam.—Polystyrene foam is characterized by excellent resistance to bacterial and mold growth. Water absorption is almost nil. Even when polystyrene foam is immersed in water for prolonged period of time it will not absorb more than 1 to 2% moisture. The material is inert, neutral and only attacked by certain oils containing aromatic constituents. It is lightweight, nonabrasive and has excellent cushioning properties. An outstanding property of polystyrene foam is its low thermal conductivity, with a K value at a density of 0.22, 2 lb per cu ft. All these properties point toward its large-scale use in cushioning and thermal insulation.

Open-cell Polystyrene Foam.—Open-cell foam is made by the extrusion of specially formulated polystyrene pellets which contain a blowing agent and a nucleating system. The blowing agent generates the pressure needed to create the cell structure. Commonly used blowing agents are n-pentane, isopentane or methyl chloride. These materials are volatile and low boiling. The nucleating agent serves to control the cell size. By utilizing different nucleating systems varying cell sizes are obtained. A commonly used system consists of a mixture of citric acid and sodium bicarbonate. The polystyrene pellets are prepared by injecting both the volatile blowing agent and nucleating system into molten polystyrene which is kept under pressure of 500 to 1,000 psi. Extrusion follows with the use of a conventional extruder. The screw should have a compression ratio of 2:1 and a length/diameter ratio of 20 to 24:1. Films below 0.050 in. in gauge may be produced as a free-blown bubble. Beyond this gauge a sizing mandrel should be used. The tubular sheet is then converged into a lay-flat tube and slit or centerfolded as required.

Closed-cell Polystyrene Foam.—Closed-cell foam is the most widely used type for packaging. It was first developed in Germany in 1950 and is now used extensively for molded packages. Production consists of these essential steps: pre-expansion (prefoaming), conditioning, and molding.

The raw material used is made from spherical beads of polystyrene which incorporate a low boiling point hydrocarbon such as petroleum ether or pentane. Various bead sizes are available with the smaller sizes used for the thinner molded sections (3/16 to 1/4 in.). The raw material is expanded to the bulk density required in the final molding by heating. This causes the hydrocarbon to vaporize, softens the polystyrene and creates a fully closed microcellular structure within each bead. Steam is generally used and the beads are expanded to about 25 times their original size. The process can be conducted on either a batch or continuous basis. For heavier density foams (i.e., 3 to 4 lb per cu ft) heating by hot water affords more accurate density control.

After expansion the prefoam beads are conditioned in ventilated silos using warm air. By this stabilization technique the air diffuses into the prefoam granule to fill the vacuum created within the individual cells on cooling.

In the final molding operation the beads are transferred to the mold. They are then expanded further and fused together by steam injection through openings in the wall into the cavity. The beads soften, expand and pack tightly in the mold so that they stick together in a solid mass. Cooling follows by circulating water through the mold jacket. This cooling time is the longest part of the cycle since the polystyrene foam has excellent insulating properties. If the mold is opened too soon the internal pressure in the molding causes it to bulge and distort. A 1/4 in. wall will cool in about 2 min; heavier sections require longer times.

There has been considerable development in refining the technique of manufacturing closed-cell polystyrene foam. Improved cooling methods, mold design and material modifications are now available. Large presses can now use multicavity molds with up to 100 cavities. Molds are made of cast aluminum or, for longer runs, of phosphor bronze. Fully automatic machines are also available which inject the pre-expanded beads into the preheated mold. Steam is then admitted, and cooling then ejection of the finished piece follow. Moldings are dried before delivery to the package user. It is interesting to note that there has also been development of a molding technique which uses high-frequency heating instead of steam. Rapid cycle times are possible with direct production of dry moldings. The mold cast is also reduced since plastic molds can be used.

The production costs of running a foam line are significantly cheaper than an injection-molded operation. Molds can run as low as $500 for some small machines and tool changes can be made quite rapidly. A large multicavity mold can cost as much as $4,000. The finished product averages about three times the raw material cost depending upon the length of the run.

An additional grade of polystyrene foam is flexible foam. This is produced by passing slabs of expanded polystyrene between rollers and compressing it to about one-half its original thickness. The closed cells are ruptured and flexibility and compressability increase.

Disposability Factors.—Polystyrene foam packages, i.e., cups and containers, can be disposed of in any of the solid waste disposal systems now used. Since they are not generally biodegradable the foam provides a stable and inert element in sanitary landfills. This is a definite asset because biodegradable materials can rot and cause problems of air and water pollution in a landfill. In incineration polystyrene foam burns yielding CO, CO_2 and water. Its aromatic structure causes an additional liberation of carbon black. The black smoke caused by polystyrene burning contains the carbon black, which remains in a modern incinerator and is consumed with other particulates. This prevents carbon black from becoming an air pollutant.

Rubber Hydrochloride

Known since 1805 and first synthesized in 1859, rubber hydrochloride film was originally introduced commercially by Goodyear in 1934. This date was a milestone in the flexible packaging industry since it marked the development of the first noncellulosic, transparent thermoplastic film. Pliofilm (Goodyear Tire and Rubber Co. registered trademark) was extensively used as a heat-sealing medium for flexible packages until polyethyelene captured a major share of the market. It is still used in fairly large quantities for cheese, meat and other packaging applications. Rubber hydrochloride is produced by dissolving crepe rubber in an organic solvent (benzene). The solution is subsequently chlorinated by bubbling hydrochloric acid gas through the solution.

Film Manufacture.—The solution containing chlorinated rubber in benzene yields a rather tough and brittle polymer upon evaporation. Antioxidants and plasticizers such as dibutyl sebacate are added to the benzene solution in order to yield a flexible film. In spite of these additions the film tends to harden somewhat upon exposure to direct sunlight. Hydrochloric acid is liberated slowly from the film, since it is not completely saturated, and this may cause problems in packaging. Brittleness can also occur at high storage temperatures.

After all the additives used are dissolved in the rubber hydro-chloride-benzene solution, the material is cast onto a continuous highly polished stainless steel belt. The solvent is removed by passing through controlled heated ovens and the final film is stripped off the end section of the casting belt. Gauge control is conducted by adjusting the opening through which the solution passes prior to hitting the moving belt. Most rubber hydrochloride solutions are formulated to 10 to 20% solids and carefully filtered to remove any dirt or gel particles. The optics and overall gauge conformity are excellent in rubber hydrochloride films mainly because of the casting technique used in manufacture.

Film Properties.—Unplasticized rubber hydrochloride film is tough and inextensible. Plasticized varieties are quite elastic and soft. The material is glossy, transparent and has a brownish cast which is age-dependent. Oil and grease resistance is good (except for essential oils) as is acid and alkali resistance. Rubber hydrochloride is soluble in cyclic hydrocarbons and chlorinated solvents. This characteristic makes solvent sealing of the film with toluene a distinct possibility.

Moisture and gas transmission rates are dependent on the amount of plasticizer used during manufacture. As the amount of plasticizer used is increased, permeability to moisture and gases also increases. Three categories of film exist: least, moderate and highest plasticizer content. Each has distinct barrier characteristics.

TABLE 7.5
PLIOFILM PROPERTY CHART

Pliofilm	WVR (38°C)	CO_2 Transmission (STP)
Grade N1	11	0.026
Grade N2	15	0.033
Grade P4	27	0.17

Note: Permeabilities are given in different units than in the article text for comparison purposes.

The N types (Goodyear designation) are the least plasticized films available and are the stiffest because of their low plasticizer content. Usually containing 1 to 2 parts plasticizer, several varieties are currently available as satin or matte finish on one side. Oxygen transmission is about 4.6 (cc per 100 sq in. per 24 hr) for 120 gauge N1 film. The N2 film is slightly more plasticized and has an oxygen transmission of 10.0 for 120 gauge film. Their permeabilities to carbon dioxide is 10.8 and 30.0 respectively.

The P and F grades of rubber hydrochloride film are moderately plasticized with 4 to 12 parts plasticizer. They have somewhat higher moisture and gas transmission rates as well as increased tear and impact strength than the N types. For 120 gauge film oxygen permeation is about 12cc and 31 cm^3 per m^2 per 24 ms per 1 atm pressure difference.

The highest plasticized grades available are designated as FF, HM, SS and M. These are extremely soft, strong, extensible films having extremely high gas transmission properties, which makes them excellent for packaging foods demanding "breathability."

The film is heat-sealable at 250 to 300° F on conventional packaging machinery. When higher plasticized grades are used in its unsupported form, their inherent softness may be a problem on automatic machinery. Rubber hydrochloride can be oriented (tensilized) to yield a thin and strong film with excellent tear propagation lengthwise.

Film Conversion and Uses.—Rubber hydrochloride is printed using flexography or gravure methods. In the fabrication of a composite laminate using rubber hydrochloride as a sealing medium, it is important to consider the type of film selected for adhesive mounting to another substrate.

In laminations an N-type film is generally used in combination with paper, foil or another plastic film. The resultant material offers good product resistance coupled with wide heat-sealing properties. Extremely oily products hold up quite well in rubber hydrochloride laminates. Polyethylene extrusion coatings have taken over a major share of the rubber hydrochloride market as a thermoplastic sealing media. Although polyethylene is lower in cost, for specialized applications rubber hydrochloride often provides long-range answers.

Laminates of rubber hydrochloride to paper are often used in roasted and ground coffee which emits considerable amounts of carbon dioxide. The inner layer of rubber hydrochloride allows the gas to escape while preventing oxygen from entering the package.

Other types of rubber hydrochloride film are used in fresh meat, cheese and produce packaging. These are generally in the un-supported form and offer a tight, cling package with most products.

Polyester

In 1847 the Swedish Chemist, Berzelius, synthesized the first polyester resin. For over 75 yr this discovery lay hidden and until Carothers started his exploratory investigations at DuPont on high polymers, polyesters were not even thought of as a packaging film. In 1941 research started by Carothers was completed by Whinfield and Dickson in England. While working on the general problem of the relationship between crystallinity and the molecular structure of high polymers, Terylene was invented and the first polyester fiber became a reality. Purchased by DuPont and licensed to Imperial Chemical Industries Ltd. for manufacture in Europe, polyester film began to be manufactured domestically on a large scale in 1954. From a 1 million lb market in 1959 to over 12 million lb in 1975, polyester films are used for a variety of packaging applications.

Properties of Polyesters.—The most widely used polyester film in packaging is formed by the condensation polymerization of ethylene glycol and terephthalic acid. Mylar (DuPont), Melinex (ICI), Scotch-par (3M) and Hostaphane (Am. Hoescht) are examples of commer-

Courtesy of Imperial Chemical Industries, Ltd.
Plastics Division

FIG. 7.8. POLYESTER IN NOVEL USE
Melinex polyester film is used for document storage.

cially available polyester films. Various other polyesters can be produced by altering the two initial components. However, polyethylene terephthalate remains the generally used film.

Polyester films are characterized by exceptional strength and excellent chemical resistance. Tensile strength averages 25,000 psi and impact strength is 25 kg-cm. Both values mark it as a film with unusually high strength. Polyester film is not affected by changes in moisture, has excellent transparency and can be used over a wide temperature range from frozen food applications to boil-in-bag pouches. Polyesters are a good barrier to gases and odors and the absence of plasticizers makes them dimensionally stable. They are FDA approved and are widely used in food packaging.

Polyethylene terephthalate polymer exists in various physical forms. It may be amorphous or have various degrees of crystallinity, depending on the thermal treatment received. At its melting point (509°F) or above, rapid quenching followed by x-ray examination displays a material with minimal crystallinity. This amorphous material is stable up to the second-order transition temperature of

TABLE 7.6
PROPERTIES OF UNCOATED POLYESTER .001-IN.

Tensile strength (lb/in.)	25,000
Impact strength (kg/cm)	25
Tear strength (gm/ml)	13–80
Stiffness (Handle-O-Meter)	40
MVTR[1]	21
Oxygen transmission rate[2]	100
Oil & grease resistance	Excellent
Fold endurance (MIT Test)	230,000

[1] gm/m^2/24 hr at 100°F and 90% RH
[2] cc/m^2/24 hr at 1 atm partial pressure gradient at 20°C and 0% RH

about 176°F. Above this temperature it begins to crystallize and changes from a glasslike to a rubbery state. Crystallization increases with temperature increase until a maximum is reached at about 338°F. The complex behavior of polyethylene terephthalate makes its manufacture more difficult than polyethylene. During processing the film exists in three different forms and each type is of value to the packaging industry.

Manufacture.—Polyethylene terephthalate is a brittle opaque film in its crystalline state and weak in its amorphous state. The polymer must be biaxially oriented in order to obtain superior film properties. Due to the low viscosity of the polymer at melt temperature, polyester is difficult to produce as a blown-oriented film. Orientation is accomplished by a tentering process subsequent to slot-die extrusion.

In one technique used for manufacturing polyethylene terephthalate film, the amorphous polymer is extruded at 510°F onto a chill roll. This film is transparent, substantially amorphous and not extremely strong. When heated above 176°F it will crystallize, turn milky and become brittle. In this form the film is virtually useless and the amorphous film cannot be regained except by melting and re-extruding.

Following extrusion, tentering of the amorphous film is done to improve its physical and chemical properties. Since polyethylene terephthalate is a crystal-forming polymer, it would appear that orientation must immediately follow extrusion of the polymer, or else crystallization would occur under normal storage conditions. Unlike polypropylene, polyethylene terephthalate has a rather high second-order transition temperature and a minimum crystallization temperature of 212°F. Under normal storage conditions the extruded film does not crystallize and is fairly stable.

Orienting is done by either a two-stage or simultaneous tenterframe process. Both have been commercially used for producing polyester films. In the two-stage process the extruded amorphous sheet is stretched about 300% in the transverse direction at 194°F. While still being held in the tenterframe, it is heated to 300°F for 5 sec. Stretching in the machine direction follows between two nip rolls operating at different speeds. In effect, the molecules of the oriented film are now arranged in a systematic pattern so that they lie in the plane of the film. It is still substantially amorphous and, if heated to a temperature above that at which it was stretched, it will shrink. This grade is known as shrinkable polyester and is not a heat-set film. After stretching is completed the film is quenched without heat-setting.

For heat-set film production, the plane-oriented film is held under tension so that it cannot contract and is heat-set at a temperature of about 390°F. This technique crystallized the film and gives it a high degree of dimensional stability at temperatures up to the heat-setting temperatures. If the film is heated it shrinks slightly. The amount of shrinkage is dependent on the temperature, heating time and thickness of the film. At any given temperature, about 1/2 the long-term shrinkage takes place during the first 2 or 3 min. When this film is heated above the normal heat-setting temperature while held under tension, the heat-setting temperature and dimensional stability of the film are raised accordingly.

In this simultaneous technique the extruded polyester sheet with beaded edges is biaxially oriented in a tenterframe equipped with two diverging series of roller clips for holding and stretching the film.

Clamps are used to clip the sheet and they run along an increasing pitch screw in order to accelerate in the longitudinal direction simultaneously as they diverge.

Each type of polyester film produced during manufacture may be used in packaging. Amorphous film is of some interest because it can be shaped by vacuum forming. Shrinkable film is used for a variety of food applications involving a skintight wrap. Heat-set film is the most widely used polyester variation due to its high strength and dimensional stability. In addition coatings are applied to polyesters and serve to widen their packaging versatility.

Amorphous Film.—The use of amorphous film is still fairly developmental. Due to its unoriented state, amorphous polyester is somewhat brittle and is usually combined with other materials in laminations. Since vacuum forming is a possible outlet for amorphous film, activity has been directed towards this market. Shaped containers from amorphous film can be used for articles such as sutures, cooked foods and engineering components. Other outlets include drum liners, blister packs and skin packs.

Shrinkable Film.—A degree of shrink and shrink energy is obtained from the orientation process. Orientation improves tensile strength, impact strength, clarity, transparency, flexibility, gas and moisture barriers, and makes the film more difficult for tear initiation. In polyester film production this process dramatically improves the properties of the amorphous film.

The maximum shrinkage of polyester film is 25 to 45% at 158° to 248°F. It has a shrink tension of 700-1, 500 lb per in^2. This is a measure of the stress which a film exerts when it is restrained from shrinking at elevated temperatures. High tension, such as polyester exhibits, is desireable since the film becomes a structural part of the package. When using polyester shrink film, caution must be exercised in limiting both shrink temperature and time to prevent crushing or distorting of the product. Polyester film faces considerable opposition in shrink packaging from other films, for while it has excellent mechanical properties, clarity and gloss, it is an expensive material. An additional problem is its limited sealability except under special circumstances.

DuPont's 65HS Mylar is used for contour bags and tubes for shrink packaging of turkeys, poultry and smoked meats. Shrinkable grades can also be used for cheeses as well as medical accessories such as bandages.

Heat-set Film.—Most widely used in packaging is the balanced biaxially-oriented heat-set polyester film from ethylene glycol and terephthalic acid. Since polyesters are not easily heat-sealable they are usually used as a component of a lamination. In a laminate polyester adds strength as well as clarity and gloss to the composite. Applications involving the unsupported film include sterilizable suture pouches, window cartons and electrical outlets. It is a good choice for packaging sharp objects, but since it is a stiff film, it is noisy and possibly objectionable for some applications.

Heat-sealing unsupported polyester film is difficult. Using heat and pressure the heat-set film can be sealed to itself under special conditions. A very small amount of benzyl alcohol applied with a felt wick combined with heat and pressure yields an excellent seal. Temperature range used should be 335° to 385°F. Impulse sealing is also possible and this technique is especially suited for shrinkable film. These seals tend to be weak and are not liquid-proof.

Many different types of polyester laminates are used. Polyester/polyethylene is the laminate which made possible the entire boil-in-bag frozen prepared food industry. The PE-PE seal is extremely strong and the polyester film contributes overall temperature flexibility. Other laminates of interest are polyester/aluminum foil/polyethylene, combinations using polypropylene, Pliofilm and paper. Polypropylene laminates are suited for sterilizable items, Pliofilm for extremely greasy foods and aluminum foil for barrier packs and opacity.

Coated Polyester Film.—As discussed previously polyester films are not readily heat-sealed. They melt and, since they are biaxially oriented, shrink extensively at temperatures well below their melting point. By using thermoplastic coatings which melt below the shrink temperature of the film a suitable heat-sealing range can be obtained. It is interesting to note that this same principle has been applied to coated polypropylene films where orientation is also responsible for heat-sealing problems.

DuPont's M grade Mylar is an example of a two-side-coated polyester film. Polyvinylidene chloride (PVDC)-type coatings are applied and both sides are heat-sealable to each other or like surfaces. About .10 mil coating is applied on each side of the film from either an aqueous PVDC emulsion or a solvent base PVDC solution. Both systems may be used since polyesters are not sensitive to either water or solvents. These coatings also serve to improve the MVTR of the polyester film. There is also a DuPont film available (M-26) which has a .20 mil PVDC coating on each side of the film. This is used for applications involving strong crimp seals which are capable of withstanding dump loading. The bond strength of PVDC to polyester film is not particularly good. Although adequate bond levels are possible, a truly inseparable coating has not yet been developed.

Work remains to be done on proper primer solutions and/or PVDC formulations.

Other grades of polyester film available include thermoformable types used in meat packaging. When laminated to polyethylene, these materials run on automatic vacuum packaging machines over a wide range of operating speeds and temperatures. With the use of the proper primer or adhesive, excellent adhesion of polyethylene to the coated side of the polyester film is possible.

Polyvinylidene Chloride

Discovered in 1839 polyvinylidene chloride (PVDC) was synthesized in 1872. Homopolymers made in 1916 were found to have a very narrow melting range and to be very hard and brittle. In 1936 a copolymer of vinyl chloride (VC) and vinylidene chloride was prepared—the first of the "sarans." Emulsion polymerization was developed between 1940 and 1947. During World War II PVDC was used to package military goods and was first introduced as a commercial packaging material in 1946 by the Dow Chemical Company. Called Saran it was later changed to Saran Wrap to avoid the term becoming a generic name. Saran Wrap is now Dow Chemical Company's registered trade name for their PVDC copolymer film. In the U.S. packaging industry the use of saran denotes a PVDC film or coating in a specific application.

PVDC Properties.—Pure PVDC homopolymer yields a rather stiff film which is hardly suited for packaging use. When polymerized with 5 to 50% vinyl chloride, a soft, tough and relatively impermeable film results. The higher copolymers are used as coatings and provide an excellent moisture barrier. Softer copolymers yield a thick, self-supporting film which has the same barrier characteristics as the thinner coatings. Permeability is dependent on the energy required to separate sufficient atoms to form "holes" and is a measure of the cohesive energy of the polymer. PVDC is more impermeable than polyethylene even though both polymers are symmetrical. Due to the increased polarity of a PVDC molecule, it has a higher cohesive energy than polyethylene and the diffusion constant is smaller. These extremely good protective characteristics have made it an important component of many laminates. PVDC film is resistant to oils, greases and alcohols. It is attacked by tetrahydrofuran, aliphatic ethers and aromatic ketones. The tensile strength of the film is in the range of 14,000 psi and its impact strength is good under ambient conditions. The specific properties of PVDC vary according to the degree of polymerization and the proportions of copolymer present. Further modifications occur with

TABLE 7.7
MECHANICAL PROPERTIES OF VARIOUS FILMS

Film	Impact Strength (kg/cm)	Stretch (%)	Tear Strength[1]	Tensile Strength lb/sq in.
PVDC	10-15	40-80	10-20	14,000
Low-density polyethylene	7-11	225-600	100-400	2,000
Oriented polystyrene	1-5	10-40	4-20	11,000
Polyamide	4-6	250-500	50-150	14,000

[1] Elmendorf test—gm/mil
Note: All data is for 1-mil gauge film.

the addition of plasticizers such as diethyl phthalate. The use of a plasticizer allows for different degrees of softness and cling in the film.

PVDC Film Manufacture.—PVDC packaging film is made by a bubble-type process using a circular die. The copolymer resin (PVDC-VC) with plasticizer is extruded through a ring-shaped die opening at 338° F. Quenching by means of a cold-water bath held at 35° to 45° F immediately follows to amorphize the polymer. Since PVDC is a crystal-forming polymer, quenching must be done immediately after extrusion to minimize the time of exposure of the film to temperatures where the rates of crystallization is high. PVDC has a glass-transition temperature below room conditions and will crystallize at room temperatures. The film must be oriented immediately after formation. If a sufficiently cold quenching temperature is not used or an excess of time passes prior to expansion, large crystallites would form causing light scattering. After being quenched the tube is rather tacky. To prevent it from welding together at the first set of pinch rolls a lubricant such as mineral oil is used inside the tube. The film is then flattened and passed through a third set of pinch rolls located outside the water bath. All three pinch rolls operate at the same speed. As the tube passes through the third and fourth set of pinch rolls, air is introduced into the tube causing it to expand into a bubble. The degree of transverse orientation depends on the ratio between the final bubble and initial tube diameter. At the same time that the transverse directional is occuring, the fourth set of pinch rolls accelerates the tube by a factor of 3 to 4 to control longitudinal stretch. After passing through the last set of pinch rolls, the tube is collapsed and either wound as a tube or slit into a film. The finished film is the heat-shrinkable grade of PVDC copolymer used in many food applications, particularly wrapping of primal cuts of beef and of poultry.

In order to produce a heat-stable PVDC copolymer film, an additional heat-stabilization step is necessary. Subsequent to collapsing of the bubble the tube is re-expanded to its original diameter using low air pressure. It is then put through a heating tunnel, collapsed and wound. By heat treating the re-expanded bubble under air pressure, shrinkage is prevented and orientation unchanged. The growth of crystallites occurs and the film becomes heat-stabilized.

Shrink PVDC Film.—Over 10 million lb of shrinkable grade PVDC copolymer film are used annually in packaging. Orientation significantly changes the basic properties of PVDC and actually makes a useful film from a virtually useless one. Orientation improves tensile strength, flexibility, clarity, transparency and impact strength. Gas and moisture permeabilities are lowered, and tear initiation becomes difficult.

One of the first applications utilizing shrinkable grade PVDC film was in the protective wrapping of frozen poultry. The bird was placed in a PVDC bag and vacuum drawn. After sealing the bag with a metal clip it was immersed in hot water. This caused the shrinkage of the bag around the bird and an extremely tight package resulted. Known as the Cryovac process, the shrink bagging of frozen poultry was developed jointly by Dow Chemical and then Dewey and Almy Chemical Company (now the Cryovac division of W. R. Grace Co.). Various other shrink applications soon followed in the meat, cheese and frozen food areas.

PVDC shrink film exhibits a maximum shrinkage of 30 to 60%. This is the percentage of shrink from its original dimensions determined by 5-sec immersion of a sample in hot water. The shrink tension range is 50 to 150 psi. Shrink tension is the stress that a film exerts when it is restrained from shrinking at high temperatures. It is measured by using a frame containing a strain gauge in hot water in which the film strip is changed. The maximum value on the gauge is read. A 50 to 150 psi range means that a tight package results after shrinking. Higher shrink tensions might yield a crushed or distorted package if not carefully controlled during the shrink operation. PVDC shrink film will shrink at 150° to 212°F in a heat tunnel of about 203° to 235°F. It seals at 248° to 320°F yielding a tight weld seal.

Although PVDC shrink films have good impact resistance, once punctured they tear readily and are not suited for applications where nonrunning properties are desired. In addition to low internal tear resistance, heat-sealing of PVDC films is quite critical requiring very closely controlled temperatures.

Nonshrink PVDC Film.—PVDC in its heat-stabilized variety can be used in its unsupported form or in laminates. Soft copolymer films are extensively used in hand wrapping. Since their surface is rather tacky and they exhibit good cling it is easy to hand-wrap a package with PVDC. Because these soft films are limp they are extremely difficult to run on automatic packaging machinery; a less plasticized grade is necessary. Increased slip is also mandatory for high-speed production. Automatic packaging is usually restricted to fairly simple package styles. It should also be noted that as the slip increases transparency tends to decrease. Heat-sealing is also quite difficult due to its sharp melting point. In order to prevent sticking of the molten film on a heat-seal bar, either Teflon tape or coating should be used. Heavy gauge PVDC is sealed by dielectric or electronic sealing techniques.

As part of a flexible laminate PVDC can be used in its film form or as a coating on paper, film or foil. The PVDC can be adhesive laminated to many substrates using solvent adhesives. It forms an excellent barrier and is heat-sealable. Markets using PVDC laminates include drugs, bottle cap liners and various specialty foods. Representative laminates include paper/adhesive/foil/adhesive/PVDC, cellophane/adhesive/PVDC, etc. PVDC may also be coextruded with polyolefins to form either film or semirigid sheets with barrier properties.

As a coating PVDC copolymer resin is used in an organic solvent or as a water dispersion. Solvent systems are formulated by mixing PVDC resin, plasticizer, wax, destaticizers and slip agents in acetone. If high amounts of PVDC are used (90%) tetrahydrofuran is the solvent selected. In applying a PVDC solution coating care must be exercised in removing all traces of residual solvent by adequate drying. Vinyls have a tendency to "grab on" to solvent and not dry rapidly. If a PVDC coating is improperly dried residual traces of solvent may lead to blocking and/or barrier degradation.

PVDC latices are dispersions of the polymer in water. Similar additives are used in the formulation of an emulsion as in an organic solvent. The difficulty is in properly emulsifying all additives into a homogenous dispersion.

Whether to use a solvent or aqueous system depends on the substrate on which the coating is to be applied. Polypropylene is coated via an emulsion system since the film is rather solvent sensitive. Cellophane is coated with both emulsions and solutions.

Polycarbonate

Discovered in the early 1950's by both American and German researchers, polycarbonate first appeared commercially in Germany in 1959. It was introduced several years later in the United States as a film by the General Electric Company under the trade name Lexan. Since polycarbonate is a relatively new plastic, its applications in packaging have only recently emerged. It is still expensive, but with increased usage cost could become a less limiting factor.

Manufacture and Properties.—Polycarbonate resin is produced by the reaction of bisphenol A with phosgene. As a film applicable to packaging, polycarbonate is made either by solvent casting or by extrusion. Extremely hygroscopic the resin must be thoroughly dry prior to extrusion. This can be successfully accomplished by an inert gas atmosphere, shrouds and careful handling. If not properly protected against moisture pick-up, polycarbonate tends to develop streaks and air bubbles. Also resulting is a more brittle material with minimal impact resistance. During extrusion polycarbonate can be heated rather rapidly with little thermal degradation. Solvent cast film offers superior optical properties to extruded film, but at an increased cost.

Polycarbonate films have excellent clarity and heat resistance. They are very tough films with good flexibility and dimensionable stability. Characterized by very good impact strength and a high softening point, tensile strength is not affected by temperature changes. The film is chemically resistant to water, organic and inorganic acids, oils, fats and dilute acids. It is quite sensitive to attack by strongly alkaline materials such as ammonia. Polycarbonate film is also partially soluble in chlorinated hydrocarbons, ketones, esters and benzene-toluene type chemicals. The poor resistance to several chemicals is an obvious disadvantage in selecting adhesive formulations for possible lamination uses. Moisture pick-up will affect dimensions about 0.1% at an increase of moisture ranging between 0.25 to 0.5%. Permeability properties of polycarbonate are poor; water vapor and gas transmission rates are extremely high. Electrical properties are excellent accounting for its large use as a thin film in electrical condensors.

The extremely good heat resistance of polycarbonate opens up several new areas of packaging. The material is usable in the range from 140° to 265° F making it quite suitable for drug and medical kit uses involving thermal sterilization. Also because of its toughness it can be used in the packaging of hardware and various other sharp objects.

Polycarbonate film is a smooth, sparkling nonstretchable material which burns moderately producing a phenolic odor. The film is soluble in acetone and insoluble in heptane.

Thermoforming Polycarbonate.—Polycarbonate film and sheet

TABLE 7.8
PHYSICAL PROPERTIES OF POLYCARBONATE AND POLYESTER

Property	Polycarbonate	Polyester
Specific gravity	1.20	1.4
WVTR[1]	8	1.7
Oxygen transmission[2]	300	4–18
Sealing range	380–430°F	—
Tensile strength (psi)	8–9,000	17–25,000
Elongation (%)	80–100	60–110

[1] gm/100 sq in./24 hr/mil
[2] cc/100 sq in./24 hr/atm/mil

have excellent thermoforming properties. They can be drawn quite deep with excellent detail. Laminated with polyethylene as a sheet it can be utilized as the thermoformed portion of a meat package. Due to its moisture sensitivity polycarbonate sheet should be predried prior to thermoforming. A 60 mil sheet requires about 250°F for 2 h. Forming temperatures range between 350° to 420°F which are higher than most other thermoplastic sheets. A sandwich-type heater is preferred and for large parts drape and billow snapback and plug assist are used in order to eliminate chill marks.

The material can also be cold-formed. It is capable of bending, stamping, drawing, coining and rolling without any heat application. Bending is done by overbending and for a right-angle bend springback amounts to about 15 degrees. Cold rolling can be accomplished by 10% reduction passes up to a total 50% reduction.

Polycarbonates are heat-sealed at temperatures about 450°F with 2-sec dwell. Solvent sealing can be performed with methylene chloride or specially formulated adhesives. Ethylene dichloride can also be used. It can also be hot-gas welded and ultrasonically sealed. Dielectric sealing is not possible because of its very low power factor. The film is FDA approved and nontoxic.

New Applications.—Polycarbonates are used to produce various household products such as cups, jugs, beakers and baby bottles. Because of its good impact strength it can serve as a viable replacement for metal in various applications. A recent Japanese invention incorporates polycarbonate as a transparent soda water siphon. Due to the high internal pressures in soda water siphons, a glass siphon is usually covered with a wire mesh to prevent glass fractures. By the substitution of polycarbonates the wire mesh may be eliminated. Another new Japanese application involves the substitution of metal by polycarbonate as a mold for chocolate bar production. It is easier to clean and is as rigid as metal.

Polyamide

The discovery of polyamide films (nylon) was due to the massive contributions made by Carothers between 1928 and 1937. Carothers was the first to emphasize the all important concept of functionality in polymeric reactions. From his work on condensation polymerization came polyamides. Introduced by DuPont in 1938 as a fiber for woven or knitted fabrics, nylon did not become a commercial reality for packaging film applications until the late 1950's.

From minimal usage in the early 1960's to over 12 million lb annually at present, nylon films are a vital tool in the hands of a packaging engineer. Although considered a specialty film and not a member of the "big four" (paper, aluminum foil, cellophane and polyethylene), many presently available packages would not be possible without polyamide.

Definition.—Two different types of nylon films are available based on their resin manufacture. One variety is made by a condensation of mixtures of diamines and dibasic acids. These are identified by the number of carbon atoms in the diamine followed by the number of carbon atoms in the diacid. The other type is formed by a condensation of w-amino acids. Identification is made by using a single number signifying the total amount of carbon atoms in the amino acid.

Nylon 6 refers to a polyamide film made from a polymer of E-caprolactam. One material is used containing 6 carbon atoms.

Nylon 66 is formed by reacting hexamethylene diamine with adipic acid. Both materials contain 6 carbon atoms each.

Nylon 610 is made by reacting hexamethylene diamine with sebacic acid. The diamine has 6 carbon atoms and is numerically first.

Nylon 11 can be obtained from castor oil or from w-undecanolactam. When the naturally occuring castor oil is used an eleven carbon starting material is obtained. Presently nylon 11 is made from petroleum sources using w-undecanolactam.

Nylon 12 is formed from w-dodecanolactam which contains 12 carbon atoms.

In addition to these nylon films copolymerization introduces many additional varieties. Nylon 66 and 610 can be copolymerized yielding a film with a lower melting point than either of the homopolymers. Fillers, plasticizers, anitoxidants and stabilizers can also be used with any or all of the many types of nylon films. Over 100 different formulations are available in the production of nylon film, but most used in packaging applications consist of nylon 6, 66 and 11.

Courtesy of Standard Packing Corp.

FIG. 7.9. NYLON IN DEEP-DRAW APPLICATION

Nylon film is used extensively for deep-draw applications involving cured meats. Note the deep cavity on this machine.

Properties.—Although each variety of nylon film has its own characteristic properties certain similarities exist. Nylon films are characterized by excellent thermal stability, i.e., they are capable of withstanding steam at temperatures up to 284°F and dry heat to even higher temperatures. Low-temperature flexibility is excellent and they are resistant to alkalies and dilute acids. Strong acids and oxidizing agents react with nylon. In general nylons are highly permeable to moisture, but their permeability to oxygen and other gases is quite low when the films are dry. Odor retention is excellent and the films are tasteless, odorless and nontoxic. The tensile strength of nylon film is about equal to that of cellulose acetate. Its yield strength and elongation is exceptionally high, an important characteristic in obtaining thermoformed packages. Burst strength and flex strength are also high. Nylon films are considered to be about three times as strong as polyethylene. Nylon 66 has a higher melting point than nylon 6, is somewhat stiffer, yields at a higher stress level and creeps more slowly. Nylon 610 has a slightly higher melting point than nylon 6 coupled with improved barrier properties and dimensional stability. Nylon 11 and 12 have higher yield factors than nylon 6, and their barrier and dimensional stability properties approach those of low-density polyethylene. Most nylon packaging films in the United States are produced from nylon 6 while European films are usually nylon 11 due to its lower cost. Films from nylon 6 have higher temperature, grease and oil resistance than nylon 11 films.

Nylon Film in Poultry Packaging.—Due to its excellent elongation and ability to stretch without breaking, nylon and its laminates can be used in applications involving irregular shapes successfully. They can be used to package bony parts and irregular conformations quite readily. Price often is the deciding factor against using nylon for supermarket applications. Another area of potential growth is the specialty poultry market in boilable bags. Sauce-prepared poultry entrees can be packed in nylon/PE bags for use by consumers. Nylon/PE is currently being used as a boilable packaging material. The ability of nylon film to withstand high autoclaving temperatures makes a retortable pouch possible.

Polyvinyl Chloride

Discovered by a French chemist named Regnault in 1835, vinyl chloride was a mere laboratory curiosity until 1875. The polymer was then first observed when vinyl chloride was converted by sunlight into a strange solid material that was chemically resistant to acids and alkalis. Chemists worked many years to identify and synthesize polyvinyl chloride. In 1928 three independent groups in the United States and Germany developed a copolymer of vinyl chloride and vinyl acetate. Processing polyvinyl chloride became much simpler and many decomposition problems were eliminated. By 1948 all the standard types of polyvinyl chloride polymers, such as granular, emulsion, and paste polymers and some copolymers were available. Improvements in manufacturing processes and techniques and the increasing size of operating units led to swift price reductions. During 1950-1955 polyvinyl chloride producers throughout the world introduced many new and improved polymers and easily processed resins became available.

Vinyl films developed before World War II appeared as a commercial packaging film in the early 1950's. Union Carbide introduced vinyl cast film and Goodyear introduced Vitafilm, extruded polyvinyl chloride. Reynolds Metals Company introduced oriented polyvinyl chloride in 1958. With the development of shrink wrapping techniques and oriented shrinkable vinyl films on the market, a new industry was born. Production of polyvinyl chloride film steadily increased over the years. In 1975 about 150 million lb of polyvinyl chloride film were used for packaging applications.

Properties of Polyvinyl Chloride Films.—Polyvinyl chloride films can be endowed with a broad range of properties depending on formulation and method of manufacture. It is inert in its chemical behavior and resistant to oils and fats. The pure polymer is a tough material which is translucent and slightly brownish. By adding plasticizers (such as phthalates or aryl phosphates) a wide range of products, from rigid sheets to flexible films, can be obtained. Polyvinyl chloride film softens above 160°F and melts at about 356°F. It may be heat-sealed and is easily printed. The unplasticized film has the same WVTR as polyethylene but is a more efficient barrier to oxygen and other gases. Additional properties of polyvinyl chloride resins are their excellent electrical, low-temperature and abrasion-resistance properties. They are also self-extinguishing when exposed to a flame.

Courtesy of American Viscose Div., FMC Corp.

FIG. 7.10. SOFT FILM POULTRY WRAP

Stretch PVC film is used for this poultry package.

One of the most important properties of polyvinyl chloride films is their ability to be made into shrink film. The material is prestretched or oriented during manufacture and holds its stretched state until subsequently heated. By virtue of this realignment of molecules, tensile strength increases. Orientation may be in one direction (uniaxial) or two directions (biaxial). It may also be stretched equally or unequally. Most other flexible plastic films are capable of undergoing orientation and subsequent shrinking. The

Courtesy of Reynolds Metals Co.

FIG. 7.11. PVC SHRINK WRAP—SHRINKCASE

commerical shrink films have been developed due to their economics and functionality.

Manufacture.—Polyvinyl chloride resin can be made by either of two methods. One starts with coal and the other with petroleum. Both methods reach a stage where the end product is acetylene gas which is reacted with hydrochloric acid (HCl) and polymerized to produce the polyvinyl chloride resin.

Polyvinyl chloride can be transformed into a variety of end products because of the following properties: adaptability to plasticization to produce products which range from very soft to very rigid; high resistance to chemical and water penetration; and choice of individual colors.

To obtain these properties it is necessary to mix polyvinyl chloride resins with modifying ingredients (plasticizers, lubricants, fillers, stabilizers, etc.), but of all ingredients compounded with polyvinyl chloride in the manufacture of films, the stabilizer is by far the most important.

In blown-tube extrusion the extruded material emerges from the die in tubular form and is literally blown into a bubble which is subsequently cooled, collapsed and wound onto rolls. Resin mixtures in powder form are continuously fed to the extruder where they are heated and undergo transformation to plastic form. The action within the extruder provides pressure that forces the polyvinyl chloride mixture to flow through a specially designed die which gives it tubular form. Introduction of air pressure from the bottom of the die expands the tube into a bubble, thereby introducing shrink characteristics to the film. Gauge thickness is controlled by the die and the actual amount of orientation caused by blowing and drawing from the die.

Raw Materials for Polyvinyl Chloride Manufacture.—In addition to the basic polyvinyl chloride resin, various other raw materials are necessary to facilitate processing and to effect advantageous properties to finished films. These other raw materials, or additives, include: (1) plasticizers, (2) stabilizers, (3) lubricants, and (4) color toners.

The plasticizer, one of the other important additives, is compounded with the resin in specified amounts to control the degree of softness in the finished film. The stabilizer is used to minimize decomposition and degradation which might take place within the vinyl chloride polymer. Lubricants in the form of waxes are added in minute amounts to reduce friction within the in-process materials and to increase their flow rate. Color toners, also added in minute quantities, aid in bringing about optimal clarity in finished films. While precise figures are not available, the material quantities used in shrink film formulations generally fall within the following limits:

Material	Parts
Resin	100
Plasticizer	10 to 45
Stabilizer	½ to 3
Others	1 or less

Prior to extruding raw materials are mixed together in as nearly a homogeneous mass as possible in a stainless-steel-lined vessel known as a dry blender. In the bottom of this vessel are a series of high-speed rotary knives mounted horizontally. The speed of the knives and the heat generated from their friction against the raw materials bring about the homogeniety of the mixture. In operation the dry ingredients are charged into the blender. These include the resin, lubricant, color toner and, in many cases, the stabilizer. Stabilizers are also used in liquid form. The blender motor is started and as mixing begins liquid plasticizer may be slowly added in closely controlled amounts.

The most critical phase of dry blending is the attainment of complete dispersion of the plasticizer in the resin so that each particle of resin absorbs an equal amount of it. Failure to satisfy this condition may result in blemishes to the finished film, such as fish eyes (grayish, translucent spots) and degradation (usually in the form of yellowish spots).

After the plasticizer has been added and all materials are in the blender the mixture is allowed to agitate until it reaches a specified temperature. It is then cooled, at which point it is ready to be charged into the extruder. The mixture now has a dry, powdery texture since all liquid additives have been completely absorbed by the resin particles. A typical extruder, as used to produce shrink films, consists of three primary parts: (1) an electrically heated hard metal barrel; (2) a cone-shaped hard metal extrusion screw; and (3) a specially designed extrusion die.

The function of the barrel is to contain the extrusion screw and to maintain temperatures necessary for the successful processing of dry blends to plastics film. The barrel is divided into three processing zones—a mixing zone, a transition or fusion zone, and a metering zone. Independent processing temperatures for each are furnished by electrical heating units. Dry blends coming into the mixing zone are heated, their individual particles rubbing against one another as a result of the action of the revolving screw. The screw performs a final degree of mixing to the blend as it feeds it to the transition zone. In the transition zone, which is heated to a temperature somewhat higher than that of the mixing zone, the blend begins to melt and the enlarging body of the extrusion screw increases pressure. As the blend is fed into the final metering zone it becomes molten and complete fusion takes place. From here it extrudes through a filtering apparatus into the blown-tube die.

The blown-tube die assembly is made up of three primary parts: a bowl shaped outer shell, a similarly shaped inner shell or mandrel, and an air inlet. The flow of melted polyvinyl chloride being fed from the extruder is controlled by the presence of the mandrel within the outer shell. The melted mixture flows into the die, fills the gap between the mandrel and the outer shell, emerges from between the mandrel and the outer shell, and emerges from between the lips of the die in the form of a tube with uniform wall thickness. This tube of film extends up approximately 20 ft to the nip rolls which collapse it for subsequent finishing operations.

The amount of orientation (both lateral and longitudinal) and the final gauge of the extruded film are controlled by coordination of the extrusion flow rate, the amount of air pressure used to blow the tube, and the speed at which the nip rolls pull and collapse the blown tube for subsequent finishing operations. The distance between the die lips, i.e., the uniform width of the circular gap between the top outer edge of the mandrel and the top inner edge of the outer shell, introduces gauge profile to the extruding film. To perform this function, however, a constant temperature must be maintained throughout the die assembly at all times. Any variation in temperature will give rise to variation in the flow rate of the melt and a difference in gauge profile will result.

Lateral (cross-directional) orientation is produced by the air pressure which expands the extruding tube of film; longitudinal (machine-directional) orientation is produced by the speed at which the nip rolls pull the extruding film away from the die and over the bubble of air. To accomplish cross-directional orientation and reduce gauge, a controlled amount of air is introduced into the extruding tube through the die assembly at the beginning of the extrusion cycle. The arrangement of the nip rolls at the top end of the blown tube allows no escape of air, thus air pressure remains constant throughout the processing cycle.

The cooling system, located just above the lips of the die, prepares the film for orientation by freezing the free-flowing molecules within the melt into a fixed structure. The subsequent two-way stretching of this fixed molecular structure as the nip rolls pull the tube over the air mandrel provides the desired amount of cross-directional and machine-directional orientation. Ambient air at room temperature cools the extruding film as it passes over the air mandrel and up to the nip rolls. Final locking of the film's now-oriented molecular structure is accomplished by the nip rolls, which collapse it for subsequent finishing operations. The nip rolls are chilled to provide this final locking within the film's molecular structure.

On being pulled over the air mandrel, collapsed and cooled by the nip rolls, the tube of film is in a near-finished condition. Subsequent finishing operations will involve slitting and unfolding the tube, trimming to desired widths, other special operations, inspection, packing and shipping.

Since the extrusion process makes possible in-line production of plastic films from the introduction of raw materials all the way through inspection and packing, it is a relatively inexpensive operation and is used widely by the film industry. In addition to its economy in production (more product volume per man hour) initial investments are substantially lower than for other film-making equipment. In the processing of vinyl films, the extruder does offer some limitations to maximum film widths. However, blown extrusion provides width versatility (narrow to very wide). The extruder also offers limitations as to maximum gauges and optimal clarity. The very nature of the extrusion process makes it virtually impossible to produce films which are entirely free of shrink characteristics.

Solution casting consists of three major steps: (1) mixing and dissolving raw materials in a solvent; (2) film forming by casting a solution of raw materials onto a continuously moving stainless steel band; and (3) stripping and subsequent finishing of the as-cast film.

The first step takes place in the plant's mixing room where polyvinyl chloride resins and other raw materials are dissolved in a highly volatile solvent. Once in solution form the mixture is pumped from the mixing room to the casting machine where it is cast upon a continuously moving stainless steel band. The solution cast mix is now carried by the moving band through an oven in which a series of hot air zones evaporate the solvent and produce film in near-finished form. As the film emerges from the oven, the band carries it to a cooling area where it is stripped from the band and wound onto spools for subsequent finishing operations.

Mixing.—The mixing of raw materials for solution casting is performed in large stainless steel tanks in batches. Raw materials include: polyvinyl chloride, plasticizer, solvent, and other additives. (Solution casting does not require temperatures as high as those necessary in extrusion, therefore processing stabilizers are not required). These ingredients are introduced into the mixing tanks in closely controlled quantities, the resin, plasticizer and other additives making up from 25 to 35% of the batch and the solvent the remaining 65 to 75%. Because of the highly flammable nature of the solvent, inert gas is pumped into the mixing tank to purge out all oxygen and form a blanket over the raw materials in solution, thereby minimizing the danger of fire or explosion.

When all raw materials have been charged into the mixing tank and the batch has been completely mixed by a set of rotary blades mounted in its bottom, it is pumped through a filtering system and into a holding tank. There it is held at closely controlled temperatures and gently agitated to allow gas bubbles to escape. The solution is then pumped from the holding tank to a running tank where it is heated to the specified casting temperature. It is subsequently filtered and metered to the casting nozzle (die) of the band casting machine.

The stainless steel band upon which the polyvinyl chloride solution may be cast can be 60 in. wide and approximately 330 ft long and produces as-cast films with widths up to 56 in. Other dimensions can also be used for this process. This band, stretched around two drums which revolve to drive it, carries the film solution through the drying oven at speeds ranging from 40 to 60 ft per min.

Within the oven are hot air zones through which the in-process film solution passes. Thses zones range from 180° to 350°F, depending upon the particular formulation being processed, and are constructed to enable air ducts to draw off the vapor-laden air evaporated from the film solution for subsequent solvent recovery.

On leaving the oven the reverse side of the band is sprayed with cold water to effect initial cooling and to aid in releasing the film from the belt when it reaches the stripping area. At the stripping area of the casting machine, the in-process film passes through a blast of cold air which serves to facilitate stripping and spooling by increasing its stiffness. The stripping-off film is pulled through a series of rollers, which removes wrinkles and trims off rough edges, and is wound onto rolls. Film to be oriented is slit during the winding process into rolls.

The film's thickness or gauge profile is controlled by the slot-shaped nozzle (die) which casts the solution onto the band in conjunction with a Beta gauge. The Beta gauge, located at the stripping area of the casting machine, electronically measures the thickness of the film across the web of the band and records gauge variations. The operator reads the recorded thickness(es) and can effect increases or decreases to gauge profile at a particular point across the web by adjusting the nozzle opening at that particular point. By means of the Beta gauge and this intricate flow control system, gauge profile can be controlled as closely as ± 5%.

New Plastic Materials

XT Polymer.—Based on an acrylic copolymer, XT resin (American Cyanamid) offers excellent clarity, good impact resistance and stiffness. Although its gas and odor permeability are low XT polymer's permeability to water vapor is fairly high. Resistance to oil, grease, acid, alkali and aliphatic hydrocarbon is high. XT polymer suffers from poor resistance to aromatic and chlorinated hydrocarbons; products containing high amounts of alcohol cannot be packaged in XT. Future applications for XT polymer appear to be in the semirigid bottle and thermoformed piece fields. Its outstanding clarity permits the use of back printed labels often found in glass bottles. The resin may be injection or blow molded into bottles or thermoformed into tubs and trays. XT polymer is presently in the medium price range.

Phenoxies.—Since phenoxy films are chemically related to epoxy resins they are tough and scuff-resistant, and exhibit heat resistance superior to most other thermoplastics. Bottles, films and extrusion coatings made of phenoxy are resistant to oils and greases. Permeability is low relative to gases, moisture and odors. A major problem, however, is alcohol resistance. Phenoxy films are subject to stress-cracking with products containing more than 40% alcohol. The outstanding clarity of phenoxy films coupled with good impact resistance make semirigid blown bottles a feasible outlet. Several toiletry items have already been packaged in phenoxy bottles. Clarity and odor barrier are usually essential requirements for a successful toiletry bottle. Phenoxy bottles are also shatterproof.

The price of phenoxy resin is fairly high. Most curing type plastics are expensive due to the cost of curing systems. They are unlikely to become significantly cheaper, however, phenoxies are most useful for highly specialized applications.

Polyphenylene Oxide.—Originally developed for engineering applications, polyphenylene oxide (PPO) exhibits a very high softening point and retains its high strength up to 390° F. Mechanical strength, dimensional stability and acid resistance is good. The film is soluble in aromatic and chlorinated hydrocarbon solvents. Potential markets include rigid molded containers for pharmaceutical products. Polyphenylene oxide is also steam sterilizable. It is fairly expensive for other than highly specialized packaging applications.

Polyurethanes.—Polyurethane film is currently available for many packaging applications. The most important properties of polyurethane film are toughness and strength. It is very soft and abrasion resistant. Due to excellent oil and grease resistance it is currently being used as a component pouch for motor oil. Solvent resistance is good toward aromatic hydrocarbons and gasoline.

Fluorohalocarbons.—Characterized by excellent moisture barrier characteristics, fluorohalocarbon films are currently being used in many drug applications. The film forms into a blister-type pack capable of holding individual pills. Its clarity is excellent, however, it is fairly expensive.

Acrylonitriles.[1]—These polymers derive from acrylic acid and its esters or other derivatives. Incorporation of atomic nitrogen into the polymer has been demonstrated to produce excellent gas barrier properties by compacting the molecule. Acrylonitriles were employed during World War II in the manufacture of synthetic rubbers.

---

[1] From material supplied by Dr. Aaron L. Brody

Courtesy of B. F. Goodrich Co.
FIG. 7.12. POLYURETHANE POUCH

They are usually commercially produced by ammonia propylene or ammonia propane ammoxidation, or by nitric acid propylene cyanization.

The most significant property is their high barrier to gases, especially oxygen, carbon dioxide, and nitrogen, about an order of magnitude better than polyester and polyvinyl chloride. From a gas permeation viewpoint, the high-barrier nitrile materials appear to offer the greatest potential for packaging oxygen-sensitive products. Thus, these materials have been commercially introduced as Monsanto's LOPAC®, Vistron's BAREX 210®, DuPont's VICOBAR® and Borg Warner's CYCOPAC® for carbonated beverage packaging. All but the Monsanto product have been made available for formation into sheet for thermoforming purposes.

Published data indicate that the packaging of beer in these materials is still open to question. Beer is sensitive to oxygen, demonstrating perceptible change with an uptake of 1 to 2 ppm oxygen. Further, some of the acrylonitrile materials have exhibited insufficient water vapor barrier properties and in particular the Vistron product and so are frequently combined with other materials to create what are now generically designated as high-barrier plastics. The high-barrier acrylonitrile plastics have a density about 15% higher than high-density polyethylene, stiffness like polystyrene, and impact strength and processing properties much like polyvinyl chloride. LOPAC is an acrylonitrile/styrene with good CO^2 and oxygen varrier characteristics and fairly good water vapor barrier properties. BAREX is an acrylonitrile/acrylate with good gas barrier, but reportedly marginal water vapor barrier properties.

Recent reports have indicated that the acrylonitrile monomer might be extracted into the contained beverage. The materials are therefore being carefully studied for commercial beverage applications. Even the long-used ABS (acrylonitrile butadiene styrene) is now being questioned by FDA for its commercial safety to contain margarine. Acrylonitriles imparts brittleness and stiffness to plastics. To produce a material with adequate toughness for bottle use either the polymer has to be toughened by orientation processes or the polymer has to be mixed with rubber.

Although acrylonitrile resins can be fabricated on existing blow-molding equipment, they are as difficult to blow-mold as polyvinyl chloride due to their high melt viscosity. They often need rubber for toughness, since today's commercial blow-molding equipment does not allow orientation. In their testing Monsanto marketed LOPAC in several sizes in cooperation with Coca-Cola bottling companies. Both oriented and unoriented LOPAC bottles have been test marketed by Coca-Cola and it appears that oriented bottles will be used in their market introduction. Unoriented BAREX bottles were test marketed in 10-oz Pepsi-Cola bottles in 1970, and in 16-oz Pepsi-Cola bottles weighing 50 gm in 1972. The original Pepsi-Cola bottles were reported to be blown by conventional tube parison methods. Other soft drink manufacturers, such as Canada Dry and Seven-Up, have also had laboratory and market tests using acrylonitrile bottles. Several breweries in the United States have reported experiments with acrylonitrile bottles, but results to date have been reported to be less than satisfactory because of problems with pasteurization and flavor interactions. Anheuser-Busch, Miller, and Coors breweries are reported among those having experimented with bottle filling and shelf tests.

SOHIO (Vistron) announced in mid-1972 that they had successfully produced experimental BAREX bottles which were filled and pasteurized alongside equal capacity glass bottles on a commercial beer filling line. However, the time-temperature experience was so high as to be economically unfeasible. These BAREX bottles were specially designed for optimum geometry and wall distribution. The bottles were biaxially oriented utilizing the Heidenreich and Harbeck blow-molding cold tube parison process developed in Germany. They were annealed for approximately 1 hr in the vicinity of their glass transition temperature (225° to 275° F).

Courtesy of Vistron, Inc.

FIG. 7.13. TEST MARKET

The Pepsi-Cola Company used bottles made from Barex® 210 resin in a market test in the Las Vegas area during 1970. They found excellent consumer reception for the new plastic bottle. The test was the first time the public was offered a product in a Barex bottle.

The Heidenreich and Harbeck bottle-molding process, developed in cooperation with two other German companies, Rheinmetall and Reifenhauser, utilizes a precision extruded cold tube of plastic of a predetermined length. The process first extrudes discrete lengths of tubing. These cold cut tubes (parisons) are conveyed to one of three rotary tables where each tube is gripped in the center while the top and bottom portions are heated to the thermoplastic temperature range of the specific resin. A molding mandrel is inserted inside the tubing. Simultaneously the bottom of the tube is formed around the mandrel and a neck contour is wedged at the top. The preformed tubing (now a test-tube-like parison) is passed automatically to a second table where the body and bottom are heated to the thermoplastic temperature range of the resin being used. The tempered preform is then passed to a third or final forming table. At this station a three-part mold closes around the tempered parison and the mandrel. Blowing takes place through the mandrel. After cooling the molds are opened and the finished bottle is lifted off the carrier mandrel by a vacuum in the bottom section of the mold.

Orientation at 4:1 or 5:1 is the preferred stretching range for acrylonitrile materials. Additional stretch does not bring dramatic improvement and would be difficult to obtain in practice.

Edible and Water-soluble films

Starch Films.—Delmar Chemical Company introduced the first commercial package using a film based on high amylose corn starch. Ediflex by American Maize Products Company is a hydropropylated amylomaize starch material. While amylose does not form a true hot melt, it exhibits plastic flow when formulated with water and a plasticizer such as glycerol, and exposed to heat and pressure. In essence it is an extruded film. Ediflex is soluble in hot water with decreasing solubility in cool water. Excellent solubility characteristics are obtained at 130°F. The film swells on exposure to high moisture conditions and has an elongation of 10% with good flexing and folding properties. It is an excellent barrier to oxygen, carbon dioxide and nitrogen. Oil and grease resistance are excellent. FDA approved, the material is stable and does not oxidize.

Ediflex can be successfully printed, heat-sealed (220° to 250°F) and is able to machine on most high-speed packaging equipment. For lamination consideration must be given to the use of nonaqueous adhesives due to the sensitivity of the film. A growing market exists for the processor in the area of PVDC-coated film. Solvent systems are needed for the coating operation. A PVDC-coated material would provide excellent moisture protection plus the inherent gas barrier of the base film.

An interesting Japanese film based on starch is Oblate. It is made from rice starch to which a small amount of vegetable gum is added. The dilute paste is subsequently drum-dried at 215°F upon heated rolls. The properties of Oblate are not as versatile as Ediflex. Oblate appears thin and mottled and becomes quite brittle unless stored under controlled humidity.

Cellulose Films.—There are several films based on food grade cellulose now available. Two of the most promising ones are Edisol-M and Methocel (Polymers Films, Inc.). These films are based on alpha-cellulose that has been chemically modified to form hydroxypropylmethyl cellulose. The material is cast from an aqueous suspension and is soluble within a 32° to 130°F range. Elongation properties are excellent (65%) and it exhibits good flexing, folding and sealing properties. Oil and grease resistance are excellent. An interesting property is that solid materials and oils such as colors or flavorants can be incorporated directly into the film.

A more recent edible cellulose film is Klucel developed by Hercules Chemical Company. Based on hydroxpropyl cellulose, the film is soluble in both organic solvents and water. It exhibits full solubility in water below 104° F, has a low equilibrium moisture content and is completely nontacky. Klucel is a true thermoplastic resin. It may be extruded, injection-molded and blow-molded. The film is made by a patented extrusion process. Experiments have been conducted in the areas of injection-molding of pharmaceutical capsules, blow-molding of bottles and capabilities of running on automatic packaging machinery.

Collagen-type films.—Used often as sausage casings collagen films are edible but not water soluble. They are made from the material found in the corium layer of fresh animal skins. During the past several years, a modified collagen derivative has become available that is water soluble at temperatures of 158° F and higher. Prepared by treating collagen with a proteolytic enzyme such as papain, the material is also formulated with about 30% dry solids of a plasticizer such as glycerol. Additional materials may be added, i.e., emulsifiers, antioxidants, and flavorants. Starch may be used in order to raise the solids to a more economical level. The collagen derivative has an excellent gas and water barrier, good flexibility at low temperatures and is readily heat-sealed.

Acetylated Monoglycerides.—Used presently as coatings and dips, materials based on acetoglyceride formulations have a unique potential. They are used to protect foods from discoloration, freezer burn and dehydration and provide excellent barrier properties. They are grease and oil resistant, stable and completely edible.

Myvacet by Eastman Chemical Products is made from distilled acetylated monoglycerides and is a transparent, flexible film with a sharp melting point. It is used as a protective coating on poultry. Moisture loss of frozen poultry is reduced after a Myvacet spray and it also acts as a release agent if the poultry is fried. Coating weights average about 3 to 4% by weight. Other uses include coatings on frozen fish, nuts and meats.

Lepak consists of 60 to 70% acetylated monoglycerides with 30 to 40% cellulose acetate butyrate. It is also used as a spray or dip coating. Due to the rather high concentration of cellulose acetate butyrate, it is suggested that the material not be consumed. If eaten it will pass through undigested.

Other coatings now available based on acetoglycerides include Cozeen. This is formulated from acetoglycerides and the protein fraction of corn (zein). Antioxidants and flavorants are usually added.

Alginates and Pectins.—Alginates occur in certain species of brown kelp seaweed and are used as thickening agents in various foods. Sodium alginate and propylene glycolalginate are soluble in acid solutions. Soluble alginate films can be obtained by evaporating the soluble acid solution on supports and subsequent stripping of the film. Pectinates as calcium-alkali derivatives form tender, edible films exhibiting a slightly salty taste. Completely soluble in boiling water, they can be eaten with any food.

Water-soluble Non-edible Films.—A rather common error is the confusion between water-soluble films and edible films. Most edible films are water-soluble, however, not all water-soluble films are edible. Four basic classes of water-soluble films exist: polyvinyl alcohol, polyethylene oxide, cellulose types, and starch types. Cellulosics and starch type films have been discussed since they are both edible and water soluble.

Polyvinyl alcohol (PVA) films are sold in a number of formulations classified as PVA Standard Cast and PVA Water Soluble Film. Although PVA standard cast films are not intended for water-soluble applications, they will go into partial solution at temperatures above 160° F, but will leave residues. The water-soluble variety goes into complete solution at temperatures as low as 40° F without any traces of residue. PVA standard cast films are translucent, tough, rubbery, and have excellent resistance to greases, oils and organic solvents. Various formulations are available for different applications.

One is produced with a uniform mat finish in .001 and .0015 in. gauges. It is used extensively in low-pressure laminations employed to produce desk and counter top material as well as wall board. Originally the surface of these laminates after manufacture had to be rubbed to reduce the extremely high gloss. Now, however, by following the practice of covering the lay-up (successive layers of fiberglass and resin coatings) with PVA film prior to curing, the surface characteristics of the film is transferred during curing to the finished laminate, thereby reducing the gloss uniformly. This film coating may be left on the finished product to protect the surface and be stripped off by the end use fabricator.

Another is a general purpose film produced in gauges from .001 to .008 in. in increments of .001 in. It is used extensively in vacuum bag molding, molding release sheets, drum liners, and grease gun filler cartridge.

PVA water-soluble film is produced in a thickness of .0015 in. It is a translucent heat-sealable film which is soluble in both hot and cold water. This property of solubility makes it suitable for packaging all products which are ultimately dissolved in water. Some few of these include soaps, detergents, bleaches, dyes, insecticides, fertilizers and water-treatment chemicals. The dissolved film has proved to be an asset in laundering fabrics since it acts as a soil-suspending agent.

It should be noted that some chemicals, notably certain compounds found in bleaches and other products, are not compatible with PVA films. Therefore it is always necessary to test the film for a specific application before full-scale production packaging is begun.

Polyethylene oxide films have the best solubility characteristics of all water-soluble films. They can be oriented to provide good burst properties and will elongate up to 600% with no return. PEO films are used for soluble laundry bags and for packaging a variety of industrial additives.

Heat-shrinkable Films

The first shrinkable films were used by the French for the preservation of meat in 1936. Committed to defense against German invasion, French scientists perfected a saran-like film under the pressures of wartime. Shrinkable films have now become part of industrial life. New films have been found amenable for shrinking and their widespread use continues. Shrink packaging is found in industries as diverse as food and toys. Display packages utilize shrink packaging and they serve to accelerate consumer impulse buying. Machine parts are often packed in shrink films for added protection.

Types of Films.—Most flexible plastic films are capable of undergoing orientation and subsequent shrinking. The commercial

TABLE 7.9
COMMERCIAL SHRINK FILMS

Film	Maximum Shrinkage (%)	Shrink Temp (°F)	Sealing Temp (°F)	Tensile Strength (Psi)
Polyethylene				
conventional	30	240	350	1,700
cross-linked	75	225	400	12,000
Polypropylene	75	275	400	20,000
Polyester	30	200	200	20,000
Rubber hydrochloride	45	180	220	10,000
Polystyrene	50	240	275	10,000
Polyvinylchloride	60	225	300	13,000
Polyvinylidene chloride copolymer	45	180	280	12,000

Note: The values above are averages of both extremes and are intended to serve as a general guide toward selection of a satisfactory film.

shrink films have been developed due to their economics and functionality. Nine basic shrink films are currently available: polyethylene, cross-linked polyethylene, modified cross-linked polyethylene, polypropylene, polyester, rubber hydrochloride, polystyrene, polyvinyl chloride, and polyvinylidene chloride copolymer.

Shrink is accomplished during manufacture by stretching under controlled conditions. The molecular orientation of the film changes and it is quenched by cooling. Heating then releases the molecules to their original orientation and shape.

Films may be oriented in both the machine and transverse directions. In addition there may be varying degrees of shrink, i.e., balanced, unequal, uniaxial. A film stretched unequally may find use in easy-open applications. One direction tears with greater ease than the other direction. It is difficult to obtain a truly balanced film due to the complexities involved in orientation. All the changes in physical properties are present, however, there is no tendency for the film to distort or shrink upon the application of heat. A review of the available shrink films precedes their machinability and applications. All of the films described have been discussed above in the context of general flexible packaging. They are reviewed here in terms of shrink packaging applications only.

Polyethylene.—Three commercial versions of shrinkable polyethylene are available. Conventional polyethylene film may be stretched, however, its tensile strength and percentage of shrink are not equal to its other modifications. If polyethylene is irradiated with high energy electrons, cross-linkages occur and the resultant tensile strength is increased. There are two modifications of the irradiation product and the amount of shrink differs. The maximum amount of shrink possible for irradiated polyethylene is 70 to 80%, while conventional film will only shrink between 15 to 40%.

Polypropylene.—Polypropylene may be oriented in varying directions and degrees of stretch. Most of the properties of polypropylene are increased by orienting. Stiffness, barrier properties, and low-temperature character are significantly altered. The film is capable of attaining between 70 to 80% shrink at good machine temperatures. Oriented polypropylene is an economical shrink film as contrasted to the special grades of polyethylene.

Polyester.—Economics and limited degree of shrink are the major disadvantages in using polyesters as shrink films. While the tensile strength of oriented polyester film is greater than either oriented polyethylene or oriented polypropylene, the maximum degree of shrink rarely exceeds 35%. The film is used for certain packaging applications involving costly products.

Rubber Hydrochloride.—Pliofilm (rubber hydrochloride) is a relatively old film but its use as a shrink film is more recent. The degree of shrink is about 50% and it possesses a fairly high tensile strength. Since the base film has excellent red-meat keeping quality, it is used extensively for the packaging of irregularly shaped cuts of meat. It is also extremely grease resistant and is in an acceptable economic range.

Polystyrene.—Polystyrene is one of the most economical films available. Its greatest drawback is the lack of suitable oxygen barrier. However, it is used fairly extensively as a "breathable" film for produce packaging. It is capable of undergoing 50% shrink and does have a high tensile value. The poor low-temperature character of polystyrene is improved by stretching. This enables the film and sheet to be used in low-temperature applications. An entirely new field has been opened to polystyrene due to its ability to undergo orientation.

Polyvinyl Chloride.—Vinyl films are used extensively in the shrink industry. The degree of shrink possible is in the 70% range and applications are varied. Fairly favorable economics coupled with versatility enable it to be one of the more widely used films. Produce utilizes a significant amount of shrinkable polyvinyl chloride film, however, bundling of multicans, display units and meat are also large-volume users.

Polyvinylidene Chloride.—The earliest usable shrink film was saran and its copolymers. The cryovac process introduced in the United States in 1948 was first applied to the packaging of poultry. The film is capable of undergoing an absolute shrink value of 60% and has a soft texture. The inherent, excellent oxygen barrier properties make it a particularly valuable film. While it is not an inexpensive film, its usefulness and technology have been difficult to match.

BIBLIOGRAPHY

ANON. 1973A. Lettuce packaging film. Food Drug Packaging 29, No. 3, 16.
ANON. 1973B. New LDPE film. Mod. Packaging 46, No. 9, 116.
ANON. 1974A. Anti-blocking agents. Europlastics 47, No. 1, 21.
ANON. 1974B. PP film is porous, yet barrier to bacteria. Packaging News, p.11.
ANON. 1974C. New horizons for today's cellophane. Mod. Packaging 47, No. 4, 22-25.
ANON. 1974D. Bi-oriented polypropylene: Material of the future. Emballages 44, No. 313, 56-61.
ANON. 1975. High nitrile packaging films. Res. Disclosure No. 133, 59.
BRISTON, J. 1974. Recent advances in plastics. Converter 11, No. 4, 16-18.
GILES, D. 1973. The flexible packaging scene. Brit. Ink Maker 16, No. 1, 17-19.
HALL, C. W. 1973. Permeability of plastics. Mod. Packaging 46, No. 11, 53-57.

HINSKEN, H. 1973. Plastic film wraps for pre-cooked foods. Australian Packaging *21*, No. 9, 27-31,33.

KAUDER, O. S. and BRECKER, L. R. 1974. A new F.D.A. cleared organophosphite stabilizer for rigid vinyl chloride plastics. SPE 32nd Ann. Technol. Conf., 512-513.

McMULLEN, J. F. 1973. New cellophanes are stronger—guard flavors better than ever. Package Eng. *18*, No. 11, 60-64.

OGORKIEWECZ, R. M. 1974. Thermoplastics: Properties and Design. John Wiley & Sons, New York.

POTTS, J. E., GLENDINNING, R. A., and ACKART, N. P. 1974. The effect of chemical structure on the biodegradability of plastics. The Plastics Institute, London, pp.1-10.

SCHULZ-BODEKER, D. F. 1975. Transparent packaging films. Neue Verpack. No. 5, 609-610, 612, 614-615. (German)

SULLIVAN, R. W. 1973. Upgraded films meet today's cooking demands. Package Eng. *18*, No. 9, 76-78.

TSUJI, K. 1975. ESR study of photodegradation of plastics. Am. Chem. Soc. Org. Coatings Plastics No. 1, 167-172.

WANG, T. C. 1975. Pouch method for measuring permeability of organic vapors through plastic films. Am. Chem. Soc. Org. Coatings Plastics No. 1, 442-447.

WINNIGER, J. M. 1973. Water quenching of cellulosic films. Mod. Plastics *50*, No. 8, 66-67.

CHAPTER 8

Flexible Packaging—Aluminum Foil

Aluminum foil is defined as a solid sheet section rolled to a thickness less than .006 in. Plain aluminum foil refers to foil in either roll or sheet form that is not combined with another material such as paper, film or cloth. (The term is used synonymously with bare foil, unmounted foil, unsupported foil, free foil and nonlaminated foil.) For most converting and many packaging operations it is used in rolls.

Specifications

Aluminum foil is normally supplied in the high purity grades (1000 alloy series), but some other alloys are now being used for special types of products. Having a minimum aluminum content of 99.35%, 1235 alloy is commonly referred to as foil analysis. It is the alloy most widely used in the packaging industry. Unless there is a specific alloy requirement in the end use, 1235 should suffice. This alloy is also used in condensers and capacitors, electrical strip conductors and many other applications. 1180 alloy, which has a minimum aluminum content of 99.8%, and higher purity alloys are used primarily for various industrial applications, particularly by the electrical industry for condensers and capacitors. With a 1.0 to 1.5% manganese content to improve physical characteristics, 3003 alloy is used largely in the fabrication of formed or drawn containers. It is also used in forming various caps and closures. MD-50 alloy, newly developed with superior strength characteristics, is now used in consumer foil and is being evaluated for packaging end uses.

Aluminum foil is produced in "O" and "H" tempers. O temper is produced by subjecting the foil, after it has been rolled to the desired gauge, to controlled heat followed by controlled cooling. Foil with O temper, commonly referred to as dead soft, is the softest form available and has the lowest physical properties. It has good folding characteristics, however, and is widely used in converting operations.

H temper is produced by strain hardening the metal in the normal foil-rolling operations, with or without supplementary thermal treatments to soften the material partially. Foil is available in a variety of H tempers:

H12 and H14 are strain hardened by rolling to temper after an intermediate anneal. This improves the physical characteristics over dead-soft foil. However, controlling the physical properties of intermediate tempers is difficult and they are not recommended where definite properties to close tolerances are involved. No guarantee can be given.

H18 is fully strain hardened by rolling. Foil rolled without intermediate or final annealing is in approximately this temper. The rolling is controlled so as to meet definite physical property limits. This is considered the direct opposite of dead-soft foil.

H19 is fully strain hardened by rolling, controlled so as to give maximum physical properties, but no guarantee is given. Some fabricating end uses require physical properties greater than ordinary, so this super-hard aluminum foil was developed.

H24 is an intermediate temper strain hardened by rolling, and then partially stress relieved by annealing. Good results have been achieved with this type of foil in container applications, but physical property limits cannot be guaranteed.

Only a limited percentage of the total production of foil is presently used in hard or in the range of intermediate tempers. Most is used in the soft or O temper. Aluminum foil is generally available in widths from 52-1/2 in. down to 1/4 in. These are not necessarily limitations but are what most producers are supplying. The very narrow foil is normally used only in the electrical industry. The thickness of foil, most frequently expressed in decimal inches but sometimes in millimeters, is commonly known as gauge. Within the industry aluminum foil is generally available in gauges from .0059 in. down to .00017 in. In packaging normally a gauge less than .00025 in. is not used, .00035 in. being the most common.

The yield of aluminum foil, which may be defined as the area provided by one pound of a specified gauge of foil, increases with the thinness to which foil is rolled. Yield is expressed in square inches per pound. Aluminum has the advantage over both tin and lead because it is easier to roll, can be commercially rolled thinner and provides more footage per pound. In the .00035 in. thickness, for example, 1 lb of aluminum covers 29,300 sq in., whereas 1 lb of lead covers only 6,979 sq in. or 23.0% of the area covered by aluminum. A pound of tin covers 10,689 sq in. or 36.6% of the aluminum coverage.

The basic unit of measure is a ream which consists of 500 sheets, size 24 in. x 36 in., or 432,000 sq in. The weight of metal per ream of plain unmounted aluminum foil is determined by dividing 432,000 sq in. by the yield in sq in. per lb. For example, the known yield for .00035 in.-gauge aluminum foil is 29,300 sq in. per lb; thus the weight of metal would be 432,000/29,300 or 14.74 lb per ream.

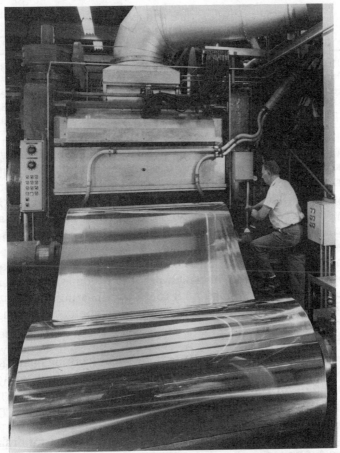

Courtesy of Reynolds Metals Co.

FIG. 8.1. FOIL IN PRODUCTION

Aluminum foil for the nation's kitchens and for packaging and other
uses rolls off a new 84-in. mill at a plant here of Reynolds Metals
Co., the country's leading producer of foil.

Ingot Preparation

Several operations are performed on the ingot prior to actual
rolling. These include scalping, homogenizing and reheating, usually
in that order. As most freshly cast ingots contain many undesirable
surface blemishes, such as small tears and oxide inclusions, it is
necessary to eliminate them by removing from 1/8 to 1/4 in. of the
surface on each side of the ingot. Large milling machines perform
this function by an operation known as scalping. This completely
cleans both faces of the ingot, makes them flat and removes surface
porosity so that only sound, clean metal remains.

TABLE 8.1
YIELD OF 1 LB. ALUMINUM FOIL IN 25 IN.-WIDE ROLLS

Gauge (In.)	Total Area (Sq in.)	Length (Ft)
.001	10,250	34.16
.0005	20,500	68.33
.00035	29,300	97.33
.00025	41,000	136.66
.00017	60,300	200.00

Homogenizing is a simple thermal treatment which relieves the internal stresses resulting from casting and uniformly distributes the various alloying constitutents throughout the ingot. Temperatures ranging from 950° to 1050°F for a period of about 24 hr are employed for homogenizing. Some alloys are so pure that this process is unnecessary, making it possible to bypass the homogenizing step.

Before rolling the ingot must be reheated to raise its temperature to about 850° or 900°F, at which point it becomes plastic and can be rolled easily. An average of 8 to 12 hr reheating time is usually required to assure uniform temperature throughout the entire ingot.

Sheet Rolling

Current practice in many plants is to begin working operations to produce sheet (a solid section rolled to a thickness range of .006 in. through .249 in.) by passing the aluminum ingot between two vertical rolls which edge roll it. This process squares up the two long edges and starts the kneading action, which breaks up the as-cast structure and eventually produces the wrought structure desired as rolling is continued.

Mills used for rolling sheet are designated by the number of vertical mill rolls in the construction. They are called 2-Hi, 3-Hi and 4-Hi. 4-Hi reversing mills are generally used for the initial or breakdown rolling of the heated ingot after it has been edge rolled (although 2-Hi mills are sometimes used). The ingot is passed back and forth between the two small hard work rolls. These are prevented from bending or springing during operation by two large back-up rolls mounted above and below. An important characteristic of rolling is that the smaller the diameter of the work rolls, the larger the degree of reduction. The pressure per unit of area transmitted from a small roll is considerably greater than that from a large roll. Reversing mills currently used by the aluminum industry are as wide as 160 in. and can accommodate ingots up to 144 in.

Next the width of the sheet is controlled by cross rolling (rolling crosswise the slab) to widen it. Thus, the first series of passes are

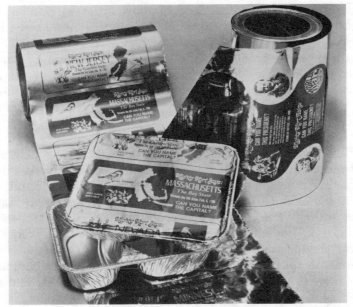

Courtesy of Reynolds Metals Co.

FIG. 8.2. FOIL IN SCHOOL LUNCH PROGRAM
Aluminum foil tuckwrap is used for school lunch feeding programs.

made by cross rolling the slab until the desired width is attained, adding a few extra inches for trim allowance. Then the slab is turned 90 degrees and rolled lengthwise for a second series of passes. As material being rolled elongates only in the direction of rolling, the increase in width during lengthwise rolling is negligible. The amount the thickness is reduced (or reduction per pass) will vary according to the alloy, width and thickness of the material being rolled. A typical reduction on relatively soft material such as 3003 is 1-1/2 in. per pass. Water-soluble lubricants are continually sprayed on the work during all hot-rolling operations. The straight-down lengthwise rolling passes are generally continued until the slab is only a few inches thick and 20 ft or more in length. At this point the irregular ends are sheared off and the edges trimmed as required. These initial or breakdown operations elongate the ingot and considerably reduce its thickness. Rolling under terrific pressure naturally causes the metal to work harden, and it is usually necessary to remove internal stresses by reheating at this stage. The slabs are then further reduced by additional hot rolling to plates about 3/4 in. thick.

After being thus reduced in thickness, the material is subjected to continuous hot rolling in a series of 4-Hi tandem mills. The term tandem means a series of mill stands erected close together and operated as a single unit. Tandem mills differ in principle from breakdown mills only in that they operate continuously in one direction. That is, a sheet passes through several mill stands at the same time, each succeeding stand going faster than the previous one to accommodate the elongating metal. As the speed of the stands increases progressively from the entrance to the exit mill, the sheet moves through them very rapidly and must be kept under high tension. Upon emerging it has usually been reduced to a thickness of about 1/8 in. This completes the hot-rolling sequence. Most of the material is coiled immediately after hot rolling, although it is sometimes allowed to cool first on a long run-out table. At this stage the coils are annealed to relieve built-up stress before cold rolling.

Cold rolling is performed on equipment very similar to that used in hot rolling. The cold-rolling process must be controlled more carefully than hot rolling because of the closer tolerances required. It requires greater power because the cold metal is harder to work. In addition the rolls of cold-mill stands are ground and polished finer than those of the hot mills. A mineral oil is utilized in cold rolling instead of the water-soluble-type oil used in hot rolling. Oil is a critical factor in cold rolling, and must be kept clean and carefully controlled for quality production. Oil serves as a coolant as well as a lubricant.

After the coils from the hot mills have cooled for a suitable period (around 36 hr) they are cold rolled in single or tandem 4-Hi mills. The thickness of the material is reduced about 50% per pass. Thus a 4-Hi two-stand tandem mill can reduce thickness by 75% in one pass through the two stands. The single-stand mills are used primarily for lighter thicknesses and usually for the narrower widths. Cold-rolling operations must be carefully planned, as a quantity of metal rolled to a certain thickness may require annealing before further rolling can be accomplished. Thus great flexibility must be provided for in scheduling both materials and equipment, as different alloys require different rolling sequences.

Sheet to be used in the rolling of foil is generally cold rolled to .023 in. thickness. (In some companies the thickness may vary up to .125 in.) It is then wound in coils ranging from 5,000 to 15,000 lb and shipped to foil plants for rerolling into foil gauges. This coiled sheet, commonly known as reroll stock, is in the as-rolled (approximately full hard) temper and has an oily surface to protect it in transportation.

Foil Rolling

Before the reroll stock can be rolled into foil it must be annealed to O temper to make it more pliable. This is necessary because cold rolling only is used; that is, the metal is not heated prior to any rolling operation. The metal is then run through a series of passes in rolling mills until the desired foil thickness (down to .00017 in.) is reached. The stock can be reduced up to a maximum of 50% during each pass in a mill's highly polished, hardened steel rolls (work rolls).

The mills used for rolling foil are similar to those used for rolling sheet but are smaller. Originally only 2-Hi mills were used for rolling foil. As the demand for foil made it desirable to roll the foil wider and faster additional strength was required and 3-Hi mills were used. Later, to get even wider rolls and faster operation, 4-Hi mills came into common usage. All modern mills are of the 4-Hi type. However, 2-Hi and 3-Hi mills are still used for rolling certain special foils, such as light gauge condenser foil and certain container stocks requiring special finishes.

Regardless of the type of mill used, the reduction in thickness is accomplished basically by the vertical pressure applied through the work rolls and by back and front tension applied to the metal through the unwind mandrel and the rewind mandrel respectively. Screws built into each side of the rolling mill provide the means of adjusting the roll pressure. In 4-Hi mills these screws are electrically controlled and actuated, as is the rewind and unwind mandrel tension. In some cases mechanical and air brakes are used on the unwind mandrel. Aluminum is rolled in every reduction in thickness in a different rolling mill. The volume of business is sufficient to make it possible to run the same reduction pass on a given mill at practically all times. Only one thickness of metal passes through the work rolls during each reducation pass until the foil reaches a thickness of generally .0015 in. Here it becomes easier to roll a better quality foil by putting two thicknesses of foil through the work rolls simultaneously. This process is called double rolling. (In some companies it is referred to as pack rolling.)

Other factors that affect reduction are the type of rolling oil used, the mill speed and the diameter of the work rolls. The normal reduction schedule for packaging foil rolled on 4-Hi mills is:

From (in.)	To (in.)
.023	.012
.012	.008
.008	.005
.005	.003
.003	.0015
.0015	.0007
.0007	.00035 (rolled double sheet)

Separating, Slitting and Spooling

After the last rolling pass is made, foil that has been rolled double is separated into single-sheet rolls on a separator. This machine has a single unwind stand and a double rewind stand. The metal passes through a series of idler rolls to maintain tension and provide a smooth sheet of foil. During the separating operation the foil can also be slit into any width desired down to about 1 in. It is also inspected and any poor quality metal is removed from the roll.

Foil is also slit into desired widths on slitters apart from the separating machines. This is particularly true in the heavier gauges (above .0015 in.) where the metal was not rolled double and also in cases where very narrow widths and more precise cuts are desired. Slitting is done by either a circular rotating knife or a razor-blade type of knife. There are four basic types of slitting. (1) Score slitting: This method utilizes a rotary knife which is a hardened steel wheel with a V edge. The knife is driven against a hardened steel roll as the web is run between the knife and the back-up roll. This action severs the metal by pressure. (2) Shear slitting: There are many types of slitters for shear cutting but two are used most frequently. One type has rotary knives mounted on two parallel shafts. As the foil is run between the knives it is cut in scissors-like fashion. The other type has a grooved shaft against which the blade is driven as the foil passes between the knife and the shaft. (3) Burst slitting: In this method the web is passed around a serrated roll and a rotary knife pierces it in the center of the serration on the back-up roll. (4) Razor-blade slitting: This type of slitting utilizes a stationary razor blade which fits into a grooved roll. The foil is slit as it passes over the roll under the fixed razor blade.

Spooling is a term used when foil is rewound to trim the edges, to remove poor quality metal and to further inspect it. During the last slitting, separating, or spooling operation the metal is wound on the core size desired.

Annealing

As mentioned previously all rolling in the manufacture of aluminum foil from reroll stock is a cold-rolling operation. As the foil is reduced in gauge it is work hardened by the repeated rolling. After an approximate 80% total reduction from the reroll gauge, it becomes full hard or H18 temper. If the end use calls for something less than this temper, the metal must be heated or annealed to relieve the internal stresses created by the work hardening. The annealing cycle, including the maximum temperature, the rate of heating, the heating time at the maximum temperature and the rate of cooling, is

Courtesy of Reynolds Metals Co.

FIG. 8.3. LAMINATE WITH BOARD

To facilitate handling convenience, a plastic carrying handle
can be attached to the new lenticular aluminum foil carton
made by Reynolds Metals Co. for Chicago's Fashion Plate
stores.

designed for the particular end use of the metal. If a temper between
dead soft (O temper) and full hard is desired, the metal is rolled to a
gauge heavier than that required and annealed to dead soft. It is then
rolled to the desired gauge and temper through one or more passes
through the rolling mill. Such a temper can also be obtained by
special annealing from full hard to the desired temper. But this
process is little used because exact physical properties are difficult to
obtain.

Surface Conditions

In the rolling and annealing operations various surface conditions
may be produced:

Dry annealed—The surface is free from an oily film. The foil is
annealed in such a way as to dissipate the residual rolling oil. The dry
surface is suitable for lacquering, printing, or coating with
water-dispersed adhesives.

TABLE 8.2

MOISTURE VAPOR TRANSMISSION RATES—
GM PER 100 SQ IN. PER 24 HR

Plain Aluminum Foil and Aluminum Foil Laminates	
.00035 in. Foil	0.37
.00035 in. Foil (Paper backed)	0.14
.0005 in. Polyethylene/.00035 in. foil (Paper backed)	0.05
.001 in. Polyethylene/.001 in. foil	0.03
Reyseal 012-W24[1]	Less than 0.05
Reyseal 502[1]	Less than 0.05
Other Packaging Materials	
.0009 in.–.0017 in. Lacquered cellophane	0.2–1.0
.001 in. Cellophane coated with .001 in. polyethylene	1.2
.001 in. Polyethylene	1.22
.001 in. Pliofilm	0.5–15.0
	(Depending on plasticizer content)
.001 in. Mylar	1.8
.001 in. Saran	0.10–0.30
.001 in. Vinyls	4.0–13.0

[1] RTM of Reynolds Metals Co. (Paper/Wax/Tissue type material)

Slick annealed—The surface has a slight film of oil. The foil is annealed in such a way as to prevent the complete dissipation of the rolling oil. The slick surface is used where better corrosion resistance than that provided by dry annealing is required or where the foil is later to be used in certain winding or machining operations.

Washed—The surface is substantially free from an oily film. During the last rolling pass the foil is solvent-flushed by a washing agent which is used instead of the usual rolling oil.

Chemically cleaned—The surface is free from all rolling oil. After the last rolling operation the foil is passed through a chemical bath. The completely dry surface is suitable for critical converting operations requiring a cleaner metal than that produced by washing. For example, chemically cleaned foil is used for container stock when a coating is to be applied.

Mechanically etched—The surface is etched by passing the foil over a mechanical brushing device after the final reduction pass. Some critical applications such as lithographic plates require this surface condition. It is not normally used in packaging.

Electrochemically etched—The surface is etched by passing the foil through an electrolytic salt bath after the final reduction pass. Foil etched by this process is used largely for electric condensers. It is not normally used in packaging.

Stearic acid-coated—The surface has a wax-like film produced by passing the foil through a bath of stearic acid in solvent solution after the final reduction pass. Foil coated with stearic acid is used in fabricating operations.

CFA (Coconut fatty acid)-coated—The surface has a slight oily film produced by passing the foil through a bath of CFA after the last rolling pass. CFA-coated foil is used for difficult forming operations where maximum lubricity is required.

Wax textured—The surface has a film of wax produced by passing the foil through a bath of wax (usually paraffin) and then chilling the wax. The operation is performed after the last rolling pass. Wax-textured foil is used for light machine applications such as bottle caps.

Finishes.—In the rolling of aluminum foil a number of finishes may be produced:

Mat one side (MIS)—One side of the foil has a diffusely reflective finish (one which scatters light and is therefore relatively dull) known as mat. The other side has a specular (bright, mirror-like) finish. MIS results from double rolling. The surfaces exposed to the bright mill rolls pick up the bright finish, whereas the inner surfaces do not receive any polishing effect. MIS is the commercial finish for foil less than .0015 in. thick. As very few end uses for thin foil call for other types of finishes, special production is seldom required.

Two sides bright (2SB)—Both sides of the foil have approximately the same specular finish, which is picked up from contact with the bright mill rolls when the single sheet rolling operation is used. 2SB is the natural commercial finish for foil above .0015 in. thick. Generally end uses for foil above .0015 in. call for this finish, so special production is seldom required.

Two sides super shiny (2SS)—Both sides of the foil have an extra bright finish which results from contact with specially polished mill rolls. 2SS is not a commercial finish and is available only on special request.

Satin finish—Both sides of the foil have a diffusely reflective surface which is produced by rolling the foil between special grit-finished mill rolls. Satin finish was developed for exclusive use on container metal.

Splicing.—Breaks in foil that occur in separating, spooling and slitting operations must usually be carefully spliced. Five types of splices standard within the aluminum industry are: (1) Foil tape: The ends of the foil are butted together, and foil tape with adhesive on one side is applied to the bottom and top of the butt joint. This type of splice is the one most frequently used. (2) Plastic tape: The webs of foil are butt or lap-joined, and a clear plastic tape is applied to either one or both sides of the joint. (3) Mechanical knurl: One web

Courtesy of Reynolds Metals Co.

FIG. 8.4. LAMINATE FOR CONVENIENCE
Aluminum foil laminated pouches are being used for individual
servings of cocoa.

of foil is placed on top of another, and the two webs are joined by passing a knurled device over the lap joint. (4) Electric weld: The webs of the foil are lap-joined, and a small electrode is touched at regular intervals across the width of the top web of foil in three evenly spaced rows. This type of splice is seldom used. (5) Ultrasonic weld: The webs of foil are lap-joined and welded together by high-frequency mechanical vibration. The welder has a movable head that welds as it travels across the joint.

All splicing with the exception of ultrasonic welding is done manually. Unless plain aluminum foil is to be processed it is next inspected and packed for shipment.

BIBLIOGRAPHY

ANON. 1973A. Aluminum foil for drinks, an important step in the coming years. Packaging *10*, No. 5, 29.
ANON. 1973B. Foil laminate protects beef cube product. Packaging News, p.39.
ANON. 1973C. New packaging trends spur alufoil growth. Paper, Film, Foil Converter *47*, No. 9, 39–40, 42.
ANON. 1973D. What happens to aluminum as a packaging material in the future? Packaging *10*, No. 10, 24–25. (Swedish)
ANON. 1974A. Barrier properties of aluminum foil. British Aluminum Foil Rollers Assoc. pp.1–25.
ANON. 1974B. Aluminum in packaging. Ind. Verpak. *4*, No. 6, 15–17. (Dutch)
ANON. 1974C. Foil is now proven to be best barrier. Packaging News, p.27.
ANON. 1975A. Iron inches into foil market—with scheme to manufacture and print in-line. Packaging News, pp.7–8.
ANON. 1975B. No foil laminate doing well in coffee tests. Packaging News, p.21.
BLAKE, A. E. 1973. Plastics and aluminum give lift to tinplate can. Packaging News, pp.22–24.

EVANS, K. A., and FENWICK, S. M. 1975. Aluminum in packaging. Sheet Metal Ind. No. 1, 28-30, 32-33.

JENSEN, A. 1975A. Aluminum material—properties and usage. Emballering No. 3, 8-9,11. (Norwegian)

JENSEN, A. 1975B. Aluminum material—quality application possibilities, Part 2. Emballering No. 4, 12-14. (Norwegian)

LANE, J. H. 1973. Foil container trends. Food Manuf. 48, No. 5, 22, 25.

LILIENBACK, K. 1974. Laminating aluminum foil with polyurethane adhesives. Verpak. Rdsch. 25, No. 7, 49-52.

MAINGUENEAU, J. M. 1972. Application of migration tests to aluminum. Ann. 1st Super Sanita 8, No. 2, 500-511.

SUMNER, J. 1973. Aluminum attractive for aerosols but steel cans cheaper. Engineer 237, No. 6138, 50.

TUNBRIDGE, T. 1973. Hot foil stamping enters new era. Europlastics 46, No. 10, 57-60.

CHAPTER 9

Flexible Packaging Material Converting

INTRODUCTION

Mankind has known and used natural plastics for thousands of years. Many materials we take for granted are natural plastics, including: wood, horn, tortoise shell, cotton, wool, silk, shellac, rosin, glue, rubber, leather and casein. All but the last three have one thing in common—they need only be gathered and very simply treated to render them useful. The latter require some additional specific chemical treatment. Although leather tanning goes back thousands of years, chemical modification of rubber and casein were discovered in 1839 and 1897, respectively.

Primitive Processing

Primitive processes used to convert natural plastic from a useless to a useful form might be broadly categorized as: (1) shaping, (2) fastening and (3) forming.

Shaping is the removal of material to attain a desired shape. An example might be the hollowing out of a wooden block to form a bowl or the sharpening of a shaft of wood to form a spear or arrow.

Fastening is the attainment of shape by combination. In search of an easier method, a shaper probably shaped some crude boards and fastened them together to make a hollow box. How did he fasten them? With pegs? With wedges? With tongue and groove? With caulking? Who can say? All of these were developed. Later came glue, metal bands, and metal nails. Another kind of fastening was the twisting of fibers into yarns, the yarns into cords, the cords into nets. Still another was the weaving or plaiting of yarns into fabrics; or the weaving or plaiting of grasses and twigs into matting or baskets.

Forming is the attainment of shape by force. Thin sections of wood can be bent by force into a new shape, but when the force is released the wood returns to its original shape. Use a cord to apply the force and you have a bow that can hurl an arrow. Apply too much force and the wood will snap. Soften the wood with heat and moisture and the bend can be severe. Hold the force until the wood dries out again and the bend is permanent. From this principle came sled runners, toboggans, bent planks for boat hulls, and bent staves for barrels.

Many primitive processes involved all three procedures. Tortoise shell was softened in boiling water, scraped clean (shaping), flattened

Much of this material was prepared by Mr. Roger C. Gribbin, Jr., Richmond, Va.

(forming), laminated together by heat and pressure (fastening), molded into a desired contour (forming), and cut and polished into a comb (shaping).

The pottery maker of ancient times learned to make stone-like hollowware by using powdered stone (clay and sand) mixed with water to make a more easily formable material which could then be fused by heat. Later, 18th and 19th century artisans learned to add binders (such as glue or shellac) to powdered plastics (such as ground horn or wood flour) and then form them in molds with heat and pressure to make buttons or picture frames.

The Chemical Revolution

The 19th century and first quarter of the 20th century ushered in a period that might be called the Chemical Revolution. Chemists seeking ways to use plastic materials like rubber and gutta percha, or to attain cheaper substitute materials for expensive silk or hard-to-get ivory, made many discoveries that led to the modern plastics and plastics converting industry.

Rubber.—Natural rubber dissolved in solvent was coated on fabrics in the early 1830's to make them waterproof. Soon a calender coater was developed. These fabrics were sticky in warm weather and awkward to use. Charles Goodyear's research in 1839 introduced vulcanizing (chemical cross-linking), and molded and coated rubber goods were a $3,000,000 business by 1850. The first extruder was developed in 1850 to apply rubber or gutta percha (a similar material) to wires for insulation. Other rubber-processing equipment developed before 1850 included the masticator. In 1934 Pliofilm made from rubber-hydrochloride was introduced. It proved to be a good barrier and formed strong heat seals, which no other available film could do.

Artificial Silk.—Prior to the development of the cotton gin cotton was expensive, though no more than oriental silk. The commonplace fabrics were made from wool, linen, or a blend called linsey-woolsey. In 1664 the scientist Hooke proposed that man ought to be able to make artificial silk by emulating the spider. In 1740 an entrepreneur knitted a pair of ladies gloves from a spider web to prove Hooke's point. In 1844 John Mercer discovered that cotton could be made stronger and more lustrous by treatment with caustic soda. Sewing shops still feature mercerized cotton sewing threads.

This prompted other chemists to examine cellulose reactions. In 1846 nitrocellulose (cellulose nitrate) was discovered which led to a variety of new products: artificial skin (collodion solution) 1848; explosives ("gun cotton," "cordite") 1850-1891; artificial leather (pyroxylin coatings on paper or fabric) 1852-1884; and artificial silk (nitrocellulose fibers) 1855-1884. In 1850 collodion was tried as a coating for glass photographic plates. In 1882 amyl acetate was used to dissolve nitrocellulose and started the modern lacquer industry. In 1887 a patent was applied for in the United States for a nitrocellulose photographic film.

Ivory Billiard Balls.—In 1868 there was a worldwide concern over possible shortages of rubber, gutta percha and ivory. One of those who sought to formulate a plastic Billiard ball was John Wesley Hyatt. He combined nitrocellulose, camphor, selected solvents, and contemporary plastics fabricating technology and founded the first celluloid business in the United States. It started as the Hyatt Billiard Ball Company then became the Albany Billiard Ball Company. They soon found that many useful items could be made from celluloid and that billiard balls were not the best. In 1868 they moved to Newark and became the Celluloid Manufacturing Company. During the next 40 years John Hyatt obtained 250 patents involving plastics products and processing machinery including the Hyatt roller bearing. Another company was founded by an engineer, Charles Borroughs, and helped develop a wide line of machinery for plastics processing including compression molders, slicers, and even a blow-molding press. Celluloid found many uses: collars, cuffs, dickeys, corset stays, shoe heel covers, spectacle frames, combs, etc. However, the advantages of celluloid and cellulose nitrate plastics were dimmed by the extreme hazard of their flammability.

Cellulose chemists continued their activities, however, and in 1891 the first cuprammonium (regenerated cellulose) rayon was patented. This was followed by the viscose (regenerated cellulose) rayon process in 1892, and in 1893 experimentation was begun on cellulose acetate filament spinning. while cellulose triacetate film was made in 1900, successful use of cellulose acetate as a rayon or a film had to wait until 1904 when acetone was found to be the best solvent. By 1910 cellulose acetate film was available for photography. In 1912 continuous casting of viscose cellulose films was perfected and the new material was called cellophane. By 1921 viscose films and cellulose acetate films were replacing celluloid in so many markets that the Florida camphor plantations were abandoned. In 1927 DuPont introduced moisture-proof cellophane (coated with nitro-cellulose), which made the film useful in food packaging as a wrapper. Soon other coatings were added to make it heat-sealable as well.

Man-made Thermoplastic Films and Coatings

Having explored the cellulosics, natural rubbers and other natural plastics rather extensively, chemists began to try to imitate the

Courtesy of E. I. DuPont De Nemours and Co.

FIG. 9.1. NYLON IN INDUSTRIAL USE

Hood latch handles on this heavy duty truck are molded of DuPont's Zytel nylon resin.

polymeric molecules starting with other monomeric materials. Work with conjugated double-bonded materials like chloroprene and butadiene led to synthetic rubbers. Work with other double-bonded materials like vinyl chloride, vinyl acetate, vinylidene chloride, styrene and acrylonitrile led to the vinyl polymers, polystyrene and ABS polymers. Still later came the polyolefins, nylons and methacrylates just in time to be of use in World War II—nylon for parachutes, plexiglass for airplane canopies and bubbles, and polyethylene for cable wrap (without which radar could not have been perfected).

Converting from Polymer to Film.—Thin sheeting materials were limited in availability to primitive civilizations. Most were achieved only by way of substantial labor. Parchments and vellums were finely scraped kidskin or lambskin. Metal sheets were hammered from rods or lumps. Translucent sheeting was made from horn by splitting it, softening with heat and moisture, and flattening the sheets with pressure. Some diving goggles have been found with eyepieces made from thin scraped and polished tortoise shell scales. Early work with shellac, animal glue and rubber established techniques for plasticizing solid materials and molding them into desired shapes or extruding them into tubing. One of the earliest methods for making thin sheeting from celluloid imitates the making of a wood veneer. A block was cast and then sliced or shaved into thin sheets. Collodion was cast as early as 1847, and rubber was solvent-coated on fabrics in the early 1830's.

Solution Casting, Coating or Laminating.—The first continuous film-making processes involved solution casting. Casting is the term used where the product becomes an independent or free film. Coating is the term used where the film is applied to a substrate surface and remains attached. Laminating is the term used when the film is applied between two substrates and acts as an adhesive to bond them together. Solution casting or coating may be from a simple solution or from an emulsion, and there may be some chemical reactions as well.

Solution Casting.—An example of a simple solution would be the casting of a free film of cellulose acetate or polyvinyl chloride. The dry polymers are dissolved in a suitable solvent or solvent mixture, and plasticizers and other needed additives are introduced and blended (this can be done before, during, or after solution). The solution is filtered, deaerated and coated onto a continuous polished steel band. The band carries the material through a series of ovens where solvent is evaporated and the film is stripped off and wound. For smooth transparent films a polished steel drum can also be used. Where a textured surface is desired the film can be cast upon a textured substrate carrier web, or the steel belt or drum can be given a roughened surface.

Emulsion casting would be the same except the first oven would drive off the water and other volatiles. There would then be a second oven which would heat the particles of polymer to fuse them together in a continuous sheet.

The viscose process for making cellophane films is different in that it is much more complex and involves a number of chemical steps such as have been described.

Solution Coating.—Solution coating may be used to apply thick film coatings to the substrates or very thin protective or decorative lacquers. When the lacquer is applied in a pattern, particularly when pigmented and in registered multicolored patterns, it is called printing (which will be considered later). Simple overall solution coating can be accomplished in a variety of ways (as will also be discussed later). Solution coating of a substrate with a plastic film, such as a vinyl on a rubber, is usually done by knife coating or by calendering. In knife coating the solution is flowed onto the substrate by means of a pond or fountain, or it can be dipped. The excess is then doctored or metered off by a rigid knife. The knife contour is critical to the amount of coating weight. Sharp edges permit light applications; broad rounded edges permit heavy applications. Usually hard back-up rolls are used to establish a fixed gap between blade and web. Knife coating is a rather slow operation for viscous coatings.

In calender coating a quantity of solution is metered onto the web. First it is flowed through a two-roll nip coater and transferred to the upper roll of a two-roll calender. This transfers the solution to the web as it passes through the calender nip. This system is by far the best for applying heavy coating weights of highly viscose materials to paper or cloth or similar substrates.

Hot-melt Extrusion Casting, Coating or Laminating

The alternative to solution casting and coating is to melt the polymer and force the melt through a die. For the making of free films there are two basic methods, flat-die extrusion and circular-die extrusion. For applying coatings or for extrusion laminations the flat-die extruder is required.

The Screw Extruder.—Until the emergence of the melt most extruders are very similar in principle. They have an entry feed, usually a hopper, and a screw surrounded by a barrel. They are usually categorized by screw diameter and the length/diameter ratio. Both barrel and screw must be controllable by heating and/or cooling so that polymer melt temperature can be controlled. After years of experimentation in design, it was found that one screw could be used to do several things depending upon the pitch, shape and depth of the threads, and the clearance between the threads and barrel wall. The feed end of the screw accepts pellets of polymer and propels them forward into the melting section. In the melting section the heat generated by the mechanical shearing forces soon melts the polymer. The loss in volume during melting must be compensated by reducing clearance between barrel and screw. After melting, the polymer passes through a transition section for mixing and homogenizing. From there it passes into a compression section which builds up the pressure necessary to force it through the screen pack (a filter) and the die. Dies can be simple openings to permit flow into an injection mold in which case the screw itself may be reciprocating to propel a measured charge. They may be shaped to provide a specific contour of filament, rod, bar or tubing. The dies of interest are slot dies that result in thin sheet or film.

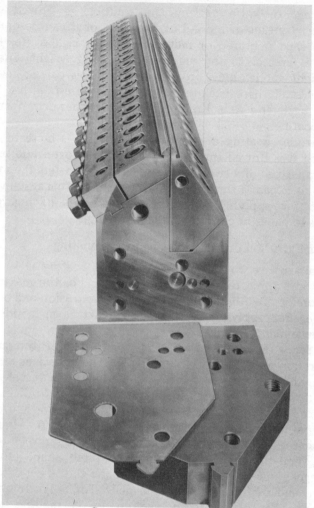

Courtesy of Black-Clawson, Co., Dilts Div.
FIG. 9.2. EXTRUSION COATING DIE
42-in. Model 300 Extrusion coating die and cap removed.

Flat-die Extruding.—The flat die accepts the melt from the extruder (usually through a center feed) and disperses it through a coat-hanger-shaped manifold through a series of segmented die lips. The tapering coat hanger shape compensates for pressure drop as the melt gets further away from the center. The die is heated and individual lip segments can be adjusted for gap opening. As the melt drops from the die it tends to shrink in width. This is called necking. Polymer properties must be carefully selected for melt strength, neck-in and rate of flow at a given temperature.

The melt can be treated in several ways: (1) it can be cast upon a continuous steel belt or a chilled steel roll and later stripped off as a free film; (2) it can be coated onto a single substrate placed between it and the chilled roll; or (3) it can be extruded between two substrates to accomplish a lamination. The distance from the die lips to the surface of the substrate is known as the air gap and can have profound influence on the characteristics of the film.

Free-film Orientation.—Films that have been flat-die extruded can be oriented by heating to a sufficient temperature (below the melt) to soften them somewhat and then by stretching in one or both web directions. They are cooled while held under tension. Lengthwise tension is accomplished by differential roll speeds. Widthwise tension is accomplished by tenter frames which grab the edges in clips mounted on diverging continuous chains. If the orienting temperature is above expected usage temperature, the film is called oriented and heat set. If the orienting temperature is low enough, the film is heat shrinkable. The application of heat above the orienting temperature will cause the film to seek to return to its original dimensions (a phenomenon known as plastic memory).

Circular-die Extrusion.—The circular-die extruder forces the melt through the lips of a circular die resulting in a tube of plastic film. Earliest types extruded upward. As the molten material rises above the die it congeals at a height known as the freeze line. At a point high enough to permit cooling the film tube passes between two pinch rolls. Usually there are several idler rolls stationed before this point to collapse the bubble prior to the pinch rolls. The bubble is created by air that is trapped inside the tube and a constant flow of air is injected to control bubble diameter. Expansion of bubble diameter above the freeze line creates cross-web orientation, and the stretching of the tube between the pinch and the freeze line creates web-direction orientation. The ring die is rotated back and forth to minimize parting lines and other gauge variations. After flattening the bubble travels to a rewind point, where it is either wound as lay-flat tubing or slit to desired single-film width. The latter may be done in-line or as a separate operation.

Water Quenching.—Early work with polyolefin extruding proved that film clarity could be improved by rapid quenching. Air cooling did not accomplish this. Several systems for extruding into water baths have been developed. For ring-die extruders this required turning upside down and bringing the bubble downward into water instead of upward into air. One of the latest systems traps water inside the bubble as well, the water inside being squeezed out by the nip of the pinch rolls.

Converting of Free Films

Free films are used in packaging as wrappers, bags and carton windows. In order to be used in this way, they may have to be rewound, slit, coated, printed, or heat-sealed. In addition they may be laminated to other substrates and then fabricated into other types of packaging.

Film Slitting—Three basic techniques are employed for slitting web materials: score, burst, and shear.

Score Cutting—Score cutting is the oldest of the three techniques. It involves the forcing of a hardened circular scoring knife against the web which is backed by a still harder platen roll. The harder and less resilient the web, the more likely will be the occurrence of particle separation or slitter dust. This is created by the differential in surface speeds of the web and the cutter blade. The web surface in contact with the platen is travelling at platen speed. The opposite or outer surface of the web is travelling faster than the platen surface. The knife at the contact point with the platen is travelling at platen speed, but as it withdraws (due to its curvature) its surface speed relative to the web and to the platen gets slower. This has caused problems in paper converting because of excessive dust. However, score cutting has many advantages. Wear can be concentrated on the knives rather than the platen, thereby reducing expensive changes. Web tensioning is no problem because the film is in contact with and supported by the platen roll. This also makes possible a contact rewind with a long span of web travel. Fast changes of slit widths are possible through the use of banks of knives controlled by pneumatic pressure. Since this can be done automatically it is not necessary that the knives be readily accessible, except for exchanging dull ones. The thinner the web the more readily it can be score cut. One possible drawback is encountered when a wide web is to be slit into many narrow widths. The accumulated pressure of many knives can deflect the platen roll. For such applications the roll must be sturdy and preferrably backed by a supporting roller.

Burst Cutting.—Burst cutting utilizes razor blades with the blade penetrating partially or totally through the web. Since there is no hard backup, web tension is critically important. Blades can be straight or rotary, oscillating or nonoscillating, and driven or nondriven. The most important difference is slitting in grooves or in air. The best materials for razor-blade slitting have the tendency to split after being partially separated. Otherwise friction and frictional heat can build up. Groove slitting provides some partial backup to the web but increases the amount of care required in positioning the blades. The grooved roll is usually driven and the blade (if straight) is oscillated. Penetration need only be minimal for parting. Slitting in air requires total penetration. The web is stretched between two idlers. However, for very light and stretchy films it is better that these be driven by means of small torque drives. Penetration requires the blade to be mounted at an angle to the web by a pivoting holder. Oscillation can only be by change of angle. This can affect friction. The advantages of razor-blade slitting are easy access to blade changing, speed of blade changing, ability to run at slow speeds, and low costs. Disadvantages include slit quality and web tensioning problems.

Shear Cutting.—Shear cutting involves parting the web by two rotary knives (male and female) that intersect at a common tangential point to act like a pair of scissors. The area of contact of the two knives varies with the depth of penetration of the male knife into the female knife, and whether or not the shafts are parallel or slightly skewed. Kiss slitting involves the passing of the web tangentially to the female knife with no wrap. Wrap slitting brings the web around the female knife, thereby receiving support from it. Kiss slitting is more suitable to heavy and more rigid materials like paperboard. Female knives are always driven. Male knives may or may not be driven. Shear slitting is more costly in initial investment and maintenance. Knife sharpening and set up is expensive. However, there is a much greater mileage from the blades. Best results on films are achieved by driven male knives with a slight over speed and with adjustable skewing to minimize web contact.

Winding.—Two basic rewinding techniques are in use: surface and center winding. A variant, center surface winding, is also considered.

Surface Winding.—A coil of web material is placed on a nondriven shaft by utilizing one or more driving rolls. These driving rolls are positioned under the coil and are called cradle rolls. A further smoothing and compression of the material is accomplished by a lay-on roll located on top of the coil. This roll may be driven or not, but it does have a controllable downward pressure. Surface winding produces dense, hard rolls with minimal web tension. It is limited to materials with relatively uniform thickness and that are relatively incompressible.

Center Winding.—The core of the coils is driven either by direct drive or by a clutch. The force required to pull the web around the roll is transmitted through all the layers. The web tension is created by this torque pulling against the unwind brake or some intermediate nip point. Tension varies in a hyberbolic function descreasing with diameter increase. Compression forces are greatest on the inside of the roll. Although it offers the advantage of using dual shafts for staggering slit rolls, center winding makes it very difficult to control roll density and formation. Rolls are either too soft and permit telescoping, or they are too hard with bulged ends and exposed gauge bands. Lay-on rolls help make better rolls, but on multiple slitting on more than one shaft these should be individual, which makes changeovers more difficult.

Center Surface Winding.—The best design for handling thin and sensitive plastic films is to combine the virtues of center winding and surface winding. Web tension is minimized by passing the web through a series of driven rolls or nips where speed relationships are controlled to deliver a taut distortion-free web to the wind-on point. The shaft of the wind-up roll is driven by a programmed torque control to maintain desired tension in the web after it is laid on the coil. The main driving force, however, is a surface drive from the backup platen roll. This also has a controllable pressure to iron out entrapped air. The speed of the platen roll should be adjustable to the preceding nips. In many applications, center surface rewinding is often superior to the other two methods.

Coating.—Coatings are applied as hot melts, extruded thermoplastic polymers, solvent solutions, aqueous solution and emulsions or suspensions. They are applied for the purpose of making the substrate smoother and more receptive to inks, to protect printed surfaces, to introduce color, or to improve functional properties such as scuff resistance, coefficient of resistance, gas barrier, heat sealability, etc. Two basic types of coating equipment systems are in use: those that apply an excessive amount and then doctor or meter off the excess, and those that meter out a predetermined amount of coating for application to the web.

Metered excess coatings are applied by means of roll, pond, fountain or dipping and the excess is removed by flexible doctor blade, rigid knife, air knife, metering squeeze rolls, or metering bar or rod.

Flexible Doctor Blade.—Flexible blade coaters are used for applying high solids clay coatings on paper and paperboard at very high speeds (several thousand feet per minute). The method may be used in line with other coatings systems or as a separate operation. The coating is applied by roller, fountain or pond system to the web, and the excess is scraped off by means of a flexible blade held in a rigid holder against a rubber-covered backup roll. Blade angle is critical and blade wear can alter results. Care must be exerted to avoid particle buildup along the blade edge which can result in scratches. Coating viscosity and blade flexibility are also important for good results.

Knife Coating.—All knife-coating systems have one thing in

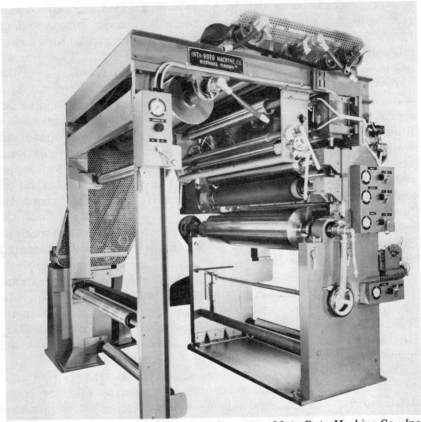

Courtesy of Inta-Roto Machine Co., Inc.

FIG. 9.3. HOT-MELT COATER

Hot-melt coatings are extensively used in flexible packaging converting.

common—a rigid steel knife. The knife edge contour is critical to the amount of coating weight. Sharp edges permit control for light application weights; broad rounded edges permit heavy coating weights. For porous webs such as woven fabrics where strike through is likely a floating knife system is used. Here no backup is provided. Pressure of the knife against the web is controlled by tension in the web itself. Up to 3 or 4 mils coating weight can be applied in one pass at rather slow speeds. Baggy edges or centers can seriously affect coating uniformity. For nonporous webs better control is available through use of backup blankets, or soft or hard backup rolls. The belt or blanket is used for delicate webs that are weak or may tend to stretch. Coatings of medium viscosity are used and light to medium coating weights can be achieved. Soft backup rolls are used where coating weight is controlled by blade pressure. Hard backup rolls are used with a fixed gap between blade and web. This system is adaptable to medium- and high-viscosity coatings run at slow speed, and will apply a wide range of coating weights.

Air Knife Coating.—Air knife coaters can use fountain or roller application of low viscosity, low to medium solids coatings at speeds as high as 1000 ft per min. A high-velocity stream of air is impinged against the web from a narrow slot opening. The air stream acts like a knife. Pressure of the air stream, angle of approach, web speed and coating properties all influence coating weight. Because solids are lower, drying can be disturbed. Air knife coating can be done on all manner of substrates, and aqueous solution, suspension, or emulsion coatings can be employed. Coatings are free of scratches and the "blade" never dulls.

Bar or Rod Coating.—The bar or rod coater employs a smooth or wire-wound rod, which is slowly rotated against the web-directional travel to smooth the coating. The coating may be applied by any method, however, usually 1- or 2-roll applicators are used. Speeds are usually slow and coatings may be very high solids, including hot melts. Smooth rods are used for lower viscosities and lower application weights. Wire-wound rods permit heavier coating weights, but coating must flow or streaks will remain. For hot-melt applications the rod may be heated.

Dip or Saturation Coating.—Coatings may also be applied by total or partial web immersion to saturate, impregnate, or 1- or 2-side coat. Surplus can be removed by blade doctoring or by metering squeeze rollers. Since this technique is usually employed for saturation, tunnel dryers are employed, weight pickup is controlled by dwell time, saturant viscosity, speeds, doctoring gap, web tensions and web wettability.

Predetermined coating weight application systems encompass several coating techniques: reverse roll, kiss coating, nip coating, direct gravure, offset gravure, spray, and curtain methods.

Reverse-roll Coating.—In reverse-roll systems the coating is picked up from a reservoir either directly by the applicator roll (3-roll system) or indirectly by means of a furnish roll (4-roll system) which transfers the coating to the applicator roll. The furnish roll is usually blade-doctored. The applicator roll is doctored by means of a metering roll which removes excess coating. Two-roll systems are also possible but are generally becoming outmoded. A doctor blade is used instead of a metering roll. The measured coating left on the applicator roll is then transferred to the web which is run between the applicator and a backup roll. The metering roll is cleaned by means of another doctoring blade. The backup roll coincides with web speed. The applicator roll runs in reverse direction and its speed and pressure against the web determine the amount of coating applied, which can be very high. Gaps and pressures of doctoring blades and rolls also affect coating weight and uniformity. Rolls are precisely ground and run in close tolerance bearings to permit gap precision. Speeds are not very high but coatings are smooth and free of scratches. The system will not level out a rough surface, so it is not well-suited for fabrics or rough boards. However, use of knife coaters in line to smooth out the latter are possible. Use of fountain application reservoirs permit use with more volatile solvents. Most reverse-roll coaters are aqueous coatings or less volatile solvent systems.

Kiss Coating.—A kiss coater is similar to a floating knife system in that the web makes contact with the applicator roll by means of web tension alone. No backup roll is used. The applicator may pick up the coating from the reservoir with little or no metering, or a second roll may be used to pick up and transfer the coating to the applicator roll. In the latter case metering is achieved by the nip between the rolls. Here again the applicator travels in reverse direction to web travel. Kiss coaters are used mostly for application tapes. It may also be used as the applicator system for air knife, blade, or bar coaters.

Nip Coating.—Nip coaters are usually 2 rolls: one hard and the other soft, or both hard. Low-viscosity coatings are applied by this method usually to both sides of the web. The coating is flooded on from above, sometimes as a spray. The web passes through the shower and retained puddle, and the nip meters the amount applied. Additional rolls may be employed for better metering or pressure control. For very thin coatings high speeds are possible. Thicker viscosities are run more slowly.

Another type of nip coater is similar to the two-roll kiss coater except it uses a rubber backup roll to hold the web against the applicator roll. The calender coater is also related in that it combines the two-roll nip coater with a calender system. Coating is flooded into a two-roll nip (both steel). Coating passes through the nip and is transferred to the upper roll of a two-roll calender, which applies the coating to the web at the calender nip. This method is used for heavy coating weights of high-viscosity materials such as vinyl coatings on paper or cloth.

Direct Gravure Coating.—The coating is picked up by impressions engraved in a metal roll. The excess is wiped off the smooth nonengraved areas by a thin doctor blade. A soft backup roll presses the web against the gravure cylinder and the coating is transferred to the web. Where viscosity permits, the coating flows together in a uniform pattern. Coating weight is controlled by viscosity, solids and engraved pattern. For the same coating heavier application requires a different gravure cylinder. Cylinder wear gradually reduces application weight. Clogging of cells can cause drastic reduction. Generally gravure coating is lightweight (less than 1 mil). This is one of the basic methods for printing. It also is used for applying lacquers, adhesives, waxes and low-melting hot melts.

Offset Gravure Coating.—Offset is very similar to direct gravure. The only difference is the engraved roll transfers the coating to the web. The advantage of the offset system is that the coating has a better chance to flow together and eliminate the line or cell pattern derived from the gravure cylinder.

Spray Coating.—This system applies low-viscosity coatings with rather nonuniform lay down. The latter is controlled somewhat by moving and rotating multiple spray heads. Losses are high. Advantages lie in being able to handle any type of substrate of any contour, and ability to use a wide range of application temperature as well as 2-phase or 2-component materials.

Curtain Coating.—A curtain coater is somewhat like an extruder. The coating material is passed through a slotted orifice under pressure. Very thin material can be applied by overflow of a weir or dam. Materials may be hot melts, lacquers or emulsions. They must be thin and free-flowing yet capable of maintaining a continuous film. Coating application weight is determined by coating flow rate, web speed and coating properties. Some curtain coaters operate at speeds of nearly 1500 ft per min. Many carton materials are curtain-coated with wax-hot melt blends.

Types of Coatings.—*Wax Coatings.*—The oldest type of coating that is still used is wax. Its greatest virtue is economy. Paraffin waxes were noted for brittleness and weak heat seals, but they provided a moisture barrier. The use of microcrystalline waxes improved these properties and the addition of polyethylenes further improved them while lowering the tendency to block. Other blends using ethylene vinyl acetate develop greater adhesion.

Wax coatings have traditionally been applied to paper, particularly glassines. These can be coated on both sides. The method of application depends upon the viscosity. Thin emulsions cannot be estrusion-coated but will be readily applied by air knife technique. When the wax is allowed to penetrate the paper by running it over a hot roll the process is called dry waxing. Wax blends for this purpose are usually almost totally paraffin. When the wax is chilled quickly before penetration it produces a smooth surface. This process is called wet waxing. Wax blends for this purpose may contain 35% microcrystalline and 5% polyethylene.

Wax coatings can be applied to plastic films and (except for some cellophanes) the film is usually a sufficient barrier without the wax. One inexpensive way to achieve a duplex structure is to wax-laminate two films.

Lacquer Coatings—Various resinous materials can be dissolved in solvent and applied to a substrate. When the solvent is driven off the resin forms a continuous coating. The most common lacquers are nitrocellulose, ethyl cellulose, polyvinyl chloride, polyvinyl acetate, and polyvinylidene chloride. Some acrylics and polyamides are also used. A very wide variety of properties can be achieved through blending of solvents, resins, plasticizers, stabilizers, and other additives. Some provide heat resistance and scuff resistance, while others are sufficiently softened by heat to act as heat-sealing materials. Lacquer coatings are applied by nearly every method

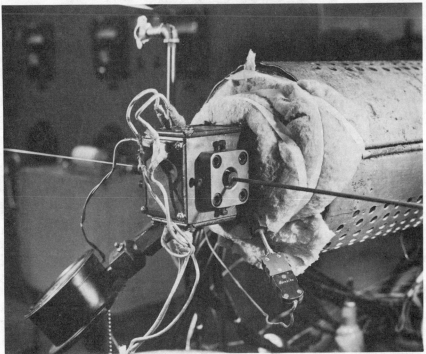

Courtesy of E. I. DuPont De Nemours and Co.

FIG. 9.4. EXTRUSION WIRE COATER

The wire enters the die at the left of center and emerges with a coating of polyethylene. The cotton-like material behind the die is additional insulating material which would ordinarily be removed. Some of these extruders are capable of coating wire at speeds up to 10,000 ft per min.

Courtesy of Inta-Roto Machine Co., Inc.

FIG. 9.5. LAMINATOR

This machine can laminate foil or paper-paperboard.

except extrusion. Pigmented lacquers can serve also as decorative coatings and for printing inks. Some solventless coatings are now being tested where the initial fluid is a monomer or partially reacted polymer plus a catalyst. After application to the substrate, a concentration of ultraviolet radiation polymerizes the resin.

Extrusion Coatings.—Most extrusion coating is done with polyethylene, polypropylene or blends. Some nylon and some polyester coating is done but price is much higher.

Laminating.—*Dry Bonding.*—Two nonporous substrates present drying problems during lamination. One of the most widely used methods for bonding two impervious webs consists of applying the adhesive to the inside face of one or both webs. The solvent is then volatilized and both webs are combined under pressure or with the aid of heat and pressure. Commercial laminators are available with in-line priming stations and the entire process is highly adaptable to in-line printing and/or coating.

Wet Bonding.—When one of the webs to be laminated is permeable the adhesive can be applied and combined while still wet. Wet-bonding is limited to applications consisting of at least one porous substrate. The process is widely used to produce laminations of paper/foil, film/paper, and paper/paper.

Thermal Lamination.—In 1961 two layers of 250-gauge polymer-coated cellophane were combined with heat and pressure. The laminate possessed outstanding durability and the process was highly economical. Thermal lamination involves passing two films through heated pressure rolls with resultant heat-sealing. Many problems were encountered during the initial stages of development and still present processing difficulties. The heat during lamination causes embrittlement of the film because moisture is removed. One method for overcoming this utilizes a specially designed laminating head which preheats the film and injects steam prior to lamination. Steam is allowed to penetrate the surface of a preheated web. The heat causes coating softening and allows for ease in moisture penetration.

Another technique employs the use of a water-nip. Water is added at the nip point and trapped air is excluded. A well of water from 1/4 to 1 in. high must be maintained in the nip between the chrome and the rubber roller.

Wax or Hot-melt Laminating.—In hot-melt laminating low molecular weight thermoplastics are used as bonding adhesives. Representative materials produced are cellophane/wax/foil, cellophane/wax/cellophane and glassine/wax/resin/glassine. The molten wax is applied to one material and brought into contact with the second web. Chilling causes the wax to set and the substrates are bonded.

Most waxes now used consist of paraffin wax blends and various microcrystalline types.

Advantages of hot-melt laminating include the lack of solvents and necessity for drying. There is also a broad versatility in weight laydown. Hot-melt laminates also provide superior WVTR and gas barrier due to the wax portion of the formulation. A limitation exists in the heat-sealing temperature. The heat-sealing temperature of the finished laminate material is generally higher than the melting point of the laminate. The net result is a movement of the wax or hot melt away from the seal area. Work is still needed in the area of high-melt viscosities and melting points.

Extrusion Laminating (Melt Laminating).—In extrusion lamination polyethylene resin is used as the adhesive between two substrates. The eventual bond is attained by means of solidified polyethylene. Conventional polyethylene extruders are used and laminating weights are usually between 7.5 to 15.0 lb per ream.

In coextrusion two or more resins are melted in individual extruders and piped into a single die. By controlling the selection of a resin and rheological properties a composite film may be obtained. There is little mixing between individual layers. The first coextruded film used in flexible packaging was introduced in 1964. It consisted of a polypropylene film sandwiched between two layers of polyethylene. In Europe nylon composites have been used for meat packaging applications. An obvious limitation inherent in coextrusion consists of the shortage of available extrudable resins with matching rheological properties. Machine and extruder design are difficult and expensive. However, once stabilized, coextrusion is a fairly simple process. Films may be produced in overall thicknesses between .00075 to .30 in.

Priming for Adhesion.—With some adhesives it is difficult to achieve a bond. It may be advantageous to apply a thin wash coat of a different material to achieve the initial bond and then apply the adhesive to the altered surface. This has been found particularly true in applying printing inks.

Printing.—Over 90% of all converted materials used in packaging are printed to some degree. Four basic printing systems are employed by converters: (1) relief, (2) planographic, (3) intaglio, (4) stencil. In the first three systems ink is applied to the printing unit and transferred to the substrate by direct contact. Stencil printing involves the preparation of a stencil so that certain areas permit ink to pass through. Other areas are designed to block ink and prevent printing.

Relief.—In relief methods the areas of the printing unit designed

Courtesy of Montedison, Italy

FIG. 9.6. FLEXO-PRINTED PASTA PACKAGES
These polypropylene packages are printed by flexographic methods.

for nonprinting are recessed below the printing areas. When ink is applied to the printing unit by a roller, only the printing areas come in contact and are receptive to ink. Transfer is accomplished by contact of the substrate only to the printing areas. Letterpress and flexography form the basic systems derived from relief processes.

Planographic.—Planographic printing is performed on a plane surface. Both printing and nonprinting areas are at the same level relative to the inking roller. Although both make contact with the roller, the nonprinting areas are treated in order to repel the ink. Lithography is a planographic method.

Intaglio.—Intaglio systems involve recessed printing areas below the nonprinting areas. After ink is applied to the entire cylinder it is scraped off by a doctor blade. When an impression is made between the cylinder and substrate the stock receives ink from the recessed cells. Gravure is an intaglio method. If gravure printing is done on a rotary press the process is called rotogravure.

Stencil.—The type of stencil printing usually employed by converters is silk screen. The nonprinting areas of the stencil are impervious to ink and the screen is fixed into a frame. Ink is placed along one edge of the screen and drawn across by a flexible squeegee. Only the printing areas allow ink to flow through and form an image.

Plastic Film Printing.—By nature some plastic films are inert and unreactive to common ink systems. Polyolefin films are notorious examples. Various treating methods are used in order to print successfully polyolefin films. These same methods can be adapted to other films if needed. Early methods involved the use of special hydrocarbon solvents coupled with polyethylene-like resins. Heat and infrared exposure then serve to allow the ink to "bite" into the film. Chemical treatments were also developed and worked successfully. These involved oxidation of the film surface. Sulphuric acid, sulphur trioxide, chromium trioxide, chlorination, fluorination, and even ozonation were evaluated with moderate success. However, chemical treatments were found to be expensive, slower and more dangerous than current methods. Full-scale commercial use of polyolefin films demanded an easier physical method. The two commonly used methods today involve surface oxidation by flame and by electronic treatment. Both yield a fully oxidized surface layer capable of wetting ink and allowing for ink permeation into the film.

Flame Treatment.—The film is passed over a chilled metal drum and the exposed surface is subjected to flames. The flame strikes the film on one surface causing a temperature gradient to occur. Only the surface is thus treated, since the chill roll cools the bulk of the film preventing any deformation. Several commercial types of burners are available, however, all involve constant length and temperature of flame. The flame is directed against the chill roll at a specified angle.

On some treating stations the rotating drum is immersed in a cooling liquid while other methods involve the use of a cooled metal shoe. It is essential that the chill roll be able to dissipate heat rapidly. If only water is used as the cooling media, temperatures between 70° to 80° F must be maintained on the circulating system. Parameters include flame temperature, distance, gas-air ratio and intensity.

The difficulty inherent in gas flame treatment is the necessity for fast rewinds. Speeds must be maintained in order to prevent the flame from scorching and wrinkling the film. It must be a fast operation and (in order to be economical) an in-tandem one. There are flame heaters in operation in-tandem with extruders. However, their inherent speeds differ. Extrusion requires slower windup speeds as contrasted with the ratio needed for flame treaters (400 to 500 ft per min). The necessity for out-of-line treating and open flames caused the plastics industry to seek still more improved methods. Most treating stations are now based on electrical discharge methods. However, flame treaters are still in limited use.

Electrical (Corona) Discharge.—Air under high-voltage gradients decomposes and becomes a gaseous conductor. It is the corona and arc from this electrical breakdown of air that is responsible for oxidizing the surface of a film. Ozone also forms and aids in the overall process. The main parts of a corona treater consist of a generator, electrode, and metal roll. The web is passed over a grounded roll. Just above it and very close is the electrode which is connected to a generator. The roll is insulated by a dielectric covering. The generated voltage can achieve grounding only by passing across the air gap and through the film and the insulative covering on the roll. Air is broken down yielding ions and ozone with a visible corona discharge. The web surface is oxidized and made printable. The air gap between the electrode and web surface has a lower dielectric breakdown than a polyolefin film. It is for this reason that the film is not damaged or excessively wrinkled. The use of a buffer layer on the grounded roll is necessary in order to prevent any discharge sparks from causing pin holes in the moving web. It is also necessary in order to assure uniform electrical discharge.

Various electrodes may be used comprising plate, shoe, glass, and vinyl types. Treating is done in-line with extrusion and controlled speeds may be maintained fairly adequately. Disadvantages include high-voltage ozone formation and static control. All of these characteristics may be controlled and the trend in the industry is toward in-line electronic treating. Stations are inexpensive and may be installed in such a manner as to treat both sides of the film simultaneously.

Measuring Treatment.—It is possible for levels of treatment to vary as well as for loss to occur. Removal of the polar groups from the treated surface may occur by physical abrasion. Wiping or brushing of the surface by rollers may induce friction and cause treatment loss. Changes in polarity may also occur. It is important for plastics processors to have at their disposal testing methods capable of measuring loss of treatment level. Several tests may be used and are not complex in nature or difficult to control: (1) wettability (contact angle method,) (2) scotch-tape test, (3) crinkle test, and (4) peel test.

Wettability.—The oxidation of the film surface involves the formation of groups capable of causing a water-wettable surface. When a drop of water or ink is placed on a treated surface it spreads over the surface. A nontreated surface will cause a water bead to occur. Quantitative measurements may be made by determining the contact angle between the liquid and plastic surface. Instruments are available in order to measure this characteristic.

Scotch-tape Test.—A lay-down of ink is made on a questionable

Courtesy of Mobil Co.

FIG. 9.7. TENSILE TESTER IN USE
The tensile testing of 0.5-mil oriented polypropylene film is being
conducted.

printed surface and allowed to dry. A length of pressure-sensitive tape is then pressed onto the surface. Removal of the tape by peeling then allows any nonadherent ink to be removed onto the tape's surface. If this occurs treatment level is marginal and no bond between ink and film will occur in production.

Crinkle Test.—In order for a film-ink surface to be acceptable no ink should flake or crack. The printed surface is folded and pinched manually for a few moments. If no flakes of ink occur the adhesion is satisfactory.

Peel Test.—A sandwich of wet ink is allowed to dry between two films. Adhesion is then tested by a peel-block on a tensile-tester.

Higher values of peel indicate excellent adhesion and satisfactory levels of treatment. One may use a tape with a low affinity for an untreated surface, but with good adhesion for a treated surface. The tape is sealed in a hand-sealer at room temperature. Peel strength is then measured by a suitable tensile-tester and values up to 430 gm per cm are obtained with films having high treatment levels.

BIBLIOGRAPHY

AIKEN, R. B. 1973. Laminations: The processes and the equipment. Flexography *18*, No. 12, 10-13, 40-41.

ANON. 1973A. Foil laminate protects beef cube product. Packaging News, p.39.

ANON. 1973B. What's happening in plastic container board? Mod. Plastics Intern. *3*, No. 10, 26-27.

ANON. 1974A. Italian paper/xps laminate looks to board market. Packaging Rev. *94*, no. 1, 51.

ANON. 1974B. Product and cost effectiveness of extruded packaging materials. Polymer Age *5*, No. 2-3, 60-61.

ANON. 1974C. Water-based laminant offers big challenge to solvents. Packaging News, p.10.

ANON. 1974D. Film firm offers more laminates more quickly. Packaging News, pp.4-5.

ANON. 1974E. 'New' laminate could find big market with biscuits. Packaging News, pp.5-6.

BOWLER, J. F. 1973. Trends in extrusion laminations for flexible packaging. Paper, Film, Foil Converter *47*, No. 10, 52.

GUILLOTTE, J. E. 1974. Coextruded films—process and properties. Polymer Plastics Technol. Eng. No. 1, 49-68.

HARTMANN, G. 1972. Film laminates pave the way for flexible packaging of fully preserved products. Gordian *72*, No. 5, 159-163.

KAGHAN, W. S., and BAKER, P. W. 1973. Laminating transparent flexible films and the water-nip process. Paper, Film, Foil Converter *47*, No. 12, 36-37.

LILIENBECK, K. 1974. Laminating aluminum foil with polyurethane adhesives. Verpak. Rdsch. *25*, No. 7, 49-52. (German)

MARION, G. J. 1974. Successful coextrusion couples material combining and equipment technology. Paper, Film, Foil Converter *48*, No. 1, 34-36.

MASON, M. E. 1975. Retort pouch appears to be convenience pack of the '80's. Packaging News, pp.17-19.

SESHAN, P. R. 1974. Flexible laminates for packaging. Packaging India *6*, No. 3, 19-25.

STOERGER, W. J. 1973. Sheets can be laminated to corrugated board with near perfect register. Boxboard Container *80*, No. 12, 37-38.

CHAPTER 10

Packaging and the Environment

INTRODUCTION

In this age of environmental concern, pollution abatement has been one of the major challenges faced by all industries. The packaging industry has not been exempt from this challenge. Millions of dollars have been spent to reduce air and water emissions from fabricating operations to meet ever more stringent state and federal regulations. However, the challenges of air and waterborne emissions are either general to all manufacturing operations or relate to specific problems affecting only a narrow part of the packaging industry. The common environmental denominator, almost unique to the packaging industry, is the challenge presented by the disposal of waste packaging materials after they have completed their basic function. Thus, the major environmental concern for the packaging industry is the solid waste problem and related subjects.

The emotional and controversial problem of litter control is clearly tied to the packaging industry. The question of energy consumption in the manufacturing, use and discarding of used packaging is a related problem. The charge of overpackaging and the call for reduced packaging through source reduction are again related to the overall problem of controlling packaging wastes. Resource recovery as an alternative or complement to solid waste disposal or to reduce use of packaging is another related topic.

The purpose of this chapter is to discuss each of these complex and interrelated topics in order to define clearly the relationships between packaging and packaging wastes, and to explore the various solutions now being pursued by private industry and by government at all levels.

SOLID WASTE

Definitions

Diverse views often turn out to be nothing more than confusion over terminology. The following definitions are those of the author and no claim is made regarding their universal acceptance. They are, however, consistent within the context of this chapter and will

This chapter has been prepared by Robert F. Testin, Ph.D.

permit the reader to understand what is being said (whether or not he agrees with it).

Solid waste—All waste materials that are not normally air or water borne. Categories: For the purpose of this chapter, three categories of solid waste will be considered—municipal[1], industrial and agricultural.

Municipal solid waste[2]—Solid waste materials resulting from households, commercial operations (hotels, restaurants, business establishments, etc.) and limited segments of the industrial community (office and cafeteria wastes, etc.). Included here are food waste (garbage), waste packaging materials, newspapers and magazines, yard wastes, street and roadside litter, etc.

Industrial solid waste—Special solid wastes from industrial operations such as scrap metals, sludges, slags, and off-grade products.

Agricultural solid waste—Solid wastes generated by agricultural operations such as animal manures, crop residues, and animal carcasses.

Litter[3]—That portion of solid waste that is scattered on streets, roadsides, parks and other private and public land rather than being placed in solid waste receptacles.

Energy—The energy required to make, transport and use the package. This quantity is generally reported in some common denominator (such as the Btu).

Overpackaging—The charge of overpackaging refers to the belief that certain packaging is superfluous or is more materials or energy consuming than some other alternative.

Source reduction—Source reduction refers to reducing solid wastes at its source (as opposed to reducing the amount of waste to

---

[1] Commercial solid wastes (from business establishments) are often distinguished from household wastes (from private homes) in many writings. In this report both are combined in the term municipal solid waste.

[2] The term municipal solid waste defines a type of material, not necessarily that generated in cities, although it is in cities that the highest concentration of this type of waste occurs. Since the main thrust of this chapter is to be on packaging wastes and their affect on solid waste disposal, and since most packaging wastes end up in municipal solid waste, the primary emphasis will be placed on the municipal solid waste category and its troublesome subcategory—street and roadside litter.

[3] The items compositing litter are similar to those found in municipal refuse, although the proportions differ. Litter generally refers to relatively small amounts of solid waste in any one location. Large concentrations of waste dumped in areas outside an authorized solid waste disposal site are generally referred to as "promiscuous dumps" rather than litter.

be disposed of through new technologies such as resource recovery). Source reduction can include the replacement of one package with an alternative that is less materials or energy consuming. In extreme cases it includes the elimination of the offending package altogether, and if necessary the product as well.

Resource recovery—The approach which considers solid waste as a resource to be mined rather than a problem to be gotten rid of. In its broadest sense resource recovery includes the reuse of waste material in any form, whether by direct salvage, conversion of waste materials to new or different products, or the use of the basic energy or chemical constituents of waste.

Amount and Composition of Solid Waste

Municipal Solid Waste.—The amount of municipal solid waste generated is subject to a great deal of confusion. Early writings in the field placed the figure at around 250 million tons per year. This figure, however, did not hold up to input/output analyses. A great number of analyses have been made on solid wastes which show that municipal and commercial solid wastes average perhaps 35% paper. Applying simple arithmetic on composition and amount would show substantially more paper in solid waste than is manufactured in the United States during a given year.

These solid waste generation figures have been revised downward to about 115 million tons per year (National Center for Resource Recovery 1974). These revised figures are a double-edged sword to the packaging industry. A landmark study prepared for the Public Health Service (Darnay and Franklin 1969) concluded that packaging wastes made up approximately 20% of municipal and commercial solid wastes. The reduced estimates of solid waste generated make the problem, *in toto*, less severe. However, packaging's contribution to the problem would be increased to about 40% because the amount of packaging produced is a known quantity, and the reduced solid waste generation figure would simply raise packaging's solid waste contribution proportionally. Table 10.1 provides a detailed breakdown of the estimated amounts of material in United States' municipal solid waste.[4]

Litter.—That portion of solid waste discarded as litter is only a minor portion (estimated at 1% by weight) of the overall solid waste

---

[4] There is still some question on this issue. The U.S. (1975) gives solid waste generation at 134.8 million tons and packaging's contribution at 35%.

TABLE 10.1
MUNICIPAL SOLID WASTES
COMPARISON OF INPUT ANALYSIS TO OUTPUT ANALYSIS

Material	Quantity per Input Analysis		Quantity per Output Analysis	
	Refuse (X 1000 tons)	% of Total	Refuse (X 1000 tons)	% of Total
Paper	38,200	33.2	42,435	36.9
Glass	10,800	9.4	9,775	8.5
Ferrous	8,900	7.7	7,475	6.5
Aluminum	630	0.55	920	0.8
Tin	66	0.06	58	0.05
Copper	355	0.3	184	0.16
Lead	30	0.03	20	0.0017
Textiles	2,900	2.5	2,185	1.9
Rubber	1,600	1.4	805	0.7
Plastics	2,200	1.9	1,265	1.1
Food, animal, plant and other wastes	49,319	42.9	49,875	43.4
Total	115,000	99.9	115,000	100.0

Source: National Center for Resource Recovery (1974).

problem. However, since litter pickup is extremely expensive on a per piece or per pound basis, and since litter is such a visual pollution problem, litter control has been one of the most popular areas of activity in the entire environmental movement. Of the waste materials comprising litter no product category has received more attention than the beer and soft drink containers. As shown in Table 10.2, beverage containers make up approximately 20% of the overall litter problem (by piece) in the United States.

Present Methods of Solid Waste Collection and Transportation

Municipal Solid Wastes.—Collection and transportation of solid wastes is extremely labor intensive. In most urban and suburban areas of the United States refuse is picked up a minimum of once a week at one of three places—at the back of the house or business establishment, in an alley behind the house or business, or at the front curb. Collection frequencies vary from once to three times a week although public health considerations require a minimum of twice weekly pickup.[5] Refuse is generally dumped manually from trash cans and placed in a standard packer truck, although some containers for commercial establishments can be unloaded automatically or hauled away container and all. The packer truck in most instances goes directly to the disposal site. In some cities material

---

[5] Twice weekly pickup interrupts the breeding cycle of flies and is therefore considered a requirement by most health authorities.

TABLE 10.2
COMPOSITION OF ROADSIDE LITTER

Item	% of Total (By number count)
Paper Items	
Newspapers or magazines	1.89
Paper packages or containers	11.52
Other	46.08
Total	59.49
Cans	
Beer	11.75
Soft drink	3.11
Food	0.64
Other	0.82
Total	16.32
Plastic Items	
Plastic packages or containers	2.57
Other	3.20
Total	5.77
Miscellaneous Items	
Auto parts and accessories (not tires)	0.83
Tires (or tire parts)	3.00
Lumber or construction items	3.97
Unclassified	4.73
Total	12.53
Bottles and Jars	
Returnable beer bottles	0.41
Nonreturnable beer bottles	2.31
Returnable soft drink bottles	1.62
Nonreturnable soft drink bottles	0.51
Wine or liquor bottles	0.64
Food bottles or jars	0.22
Other	0.17
Total	5.88
Total of all Items	99.99

Source: Finkner (1969).

may be hauled by packer truck to a transfer station, where it is loaded to larger vehicles for transportation to the disposal site.

In rural areas hauling distances preclude the possibility of door to door pickup. However, some states have instituted "green box" programs in which strategically placed large refuse containers (similar to those servicing commercial establishments) are available for the

Courtesy of Tarrant Mfg., Co.,
Saratoga Springs, N.Y.

FIG. 10.1. EXAMPLE OF ADVANCED LITTER PICKUP DEVICES

disposal of household wastes. The green boxes are serviced regularly by standard packer trucks. Uncollected rural wastes are either disposed of on site or taken by the individual to an authorized (or unauthorized) disposal site.

Litter.—The available methods of litter pickup are known to almost everyone. The park maintenance man with the shoulder sack and the pointed stick, the shopping center mall's janitor with the broom and long handled dustpan, and the highway department crews walking along the roadsides are all familiar examples of litter collection. All have one thing in common—they are extremely labor intensive and expensive. Automatic litter pickup devices have been developed. These range from small sweepers and vacuums for parking lots and sidewalks to large versions for highway and roadside use. Nevertheless, manual methods are still the rule for litter pickup across the United States. Once it has been picked up, litter is deposited in trash cans or trucks. The handling methods for transportation and disposal are the same as for municipal refuse.

Present Solid Waste Disposal Methods

There are currently only two acceptable methods of solid waste disposal in the United States: the sanitary landfill and the incinerator.

Courtesy of Tony Perrins, Private Collection (1970)

FIG. 10.2. OPEN DUMPS ARE AN UNSATISFACTORY DISPOSAL
METHOD

A sanitary landfill should not be confused with an open dump. In a sanitary landfill operation refuse is hauled to a pre-prepared site, compacted and covered with dirt each day. Care is taken to ensure that problems such as methane gas generation (from decaying organic matter) and pollution of ground water are eliminated. Tremendous progress has been made in converting the old "open dump" into sanitary landfills. However, the majority of land disposal sites still are classified as dumps.

Incineration has also been recognized as an acceptable solid waste disposal method. However an incinerator is much more costly than landfill. Air pollution regulations have required the installation of expensive air pollution control devices. Because of the relatively large capital and operating cost requirements for incinerators and lingering concern about air pollution control in terms of both feasibility and cost, the sanitary landfill is the preferred disposal method wherever

Courtesy of Reynolds Metals Co.

FIG. 10.3. IN A SANITARY LANDFILL REFUSE IS COMPACTED AND COVERED WITH DIRT EACH DAY

possible. Less than 10% of the nation's solid waste is handled through incinerators.

The Cost of Solid Waste Disposal

Good data on cost of collecting, transporting and disposing of solid waste and litter are extremely difficult to obtain. Part of this stems from the fact that in many areas only a partial accounting of the cost of solid waste handling is made public. Estimates have been made, however, that can provide an indication of the cost of solid waste disposal in this country. It is generally accepted that a sanitary landfill can be operated for about $2 per ton of refuse handled. Incineration, because of the requirement for satisfactory air pollution control, is much more expensive—ranging from $10 to $15 per ton.

Solid waste collection and transportation costs have been estimated at 50 to 90% of the total cost of solid waste handled in a community. From these data several conclusions can be drawn. First, it is obvious that a more expensive disposal method (e.g., incineration) makes sense only if available landfill sites require hauling the refuse for a substantially longer distance. Also, it can be seen that overall solid waste handling costs appear to be in the $20 to $30 per ton range. Despite the fact that litter is only 1% or so of solid wastes,

Courtesy of Tony Perrins for Reynolds
Metals Co. (1970)

FIG. 10.4. EXAMPLE OF LARGE, MODERN MUNICIPAL INCINERATOR

litter control is extremely expensive on a per piece basis because it is so labor intensive. The cost of picking up litter in the United States has been estimated at anywhere from $300 to $800 million annually. On a per piece basis estimates of litter pickup have run as high as $0.25 per item. On a national basis the total cost of solid waste management has been estimated at $3.5 million annually. Since the cost of solid waste disposal is borne by local governments, in most communities it is one of the most expensive community services.

ENERGY

The question of energy consumption is not as obviously tied to packaging as are the problems of litter and solid waste. However, the nationwide emphasis on energy conservation has made the question of energy consumption during manufacture and use one that must be dealt with when evaluating various packaging alternatives. The standard chronology of packaging's environmental problems would be: a concern about litter, a concern about solid waste and finally a charge that packaging uses too much energy. Therefore, the energy issue is one that must be dealt with in any discussion regarding packaging and the environment.

The method of dealing with the energy question was pioneered by Hannon (1973) in his analysis of beverage container alternatives. In this approach one totals all the energy required to extract raw materials from the ground, transport them to a manufacturing site, convert them into a product, and finally dispose of that product after it has completed its useful life. In order to make these calculations all energy units are converted to some common denominator, such as the British Thermal Unit (Btu). Insofar as possible the analysis takes into account both direct and indirect energy factors. An example of an indirect energy factor is that energy required to extract fuel from the ground and prepare it for final consumption. The classic example here is electrical energy where each Btu of energy in the form of electricity requires the burning of about 3 Btu of fuel due to the inefficiencies of electrical generation. For every Btu of electrical energy consumed approximately 3 Btu would be added into the product's energy consumption. All of the direct and indirect energy units are then totaled and the energy requirements of the package (or other product) can be ascertained. From this the energy savings for using another packaging alternative (or perhaps not making the package at all) can be calculated.

There is some technical question regarding whether or not this analysis can be made in a way that gives meaningful results. Inherent is the assumption that energy units are completely transferable from one form of energy to another, from one geographic location to another and from one material or product to another. This type of transferability exists, but only in a very limited sense. Yet, everyone has read in the press that a switch from one type of container to another or elimination of a product from the marketplace will save "the equivalent of X thousands of barrels of oil a day." Whether or not one agrees with this approach to energy calculations, it must be understood, because it is being used by those who wish to control the packaging industry for environmental reasons.

THE RELATIONSHIP OF PACKAGING
TO THE QUESTIONS OF LITTER, SOLID WASTE AND ENERGY

The environmental charges against the packaging industry can be categorized into a number of areas. The first and most general charge is that of overpackaging. Generally packaging's multiple roles are not appreciated by the public. Much packaging tends to be viewed as frivolous, with the belief that it could be readily eliminated or drastically reduced if society decided to take this step. A second and more specific concern is the use of non-returnable containers, particularly for beverages. Since beverages are one of the few products on the market that have retained a substantial amount of reusable packaging, this area has been attacked most frequently. A return to refillables appears to be readily available to the beverage industry by those unfamiliar with the intricacies of beverage packaging, distribution and marketing. A third concern has been the desire to force the industry, if not to returnable packages, to biodegradable ones. Thus, when a package was required and could not be reused, it would decompose naturally when discarded as solid waste or even as litter.

In summary the environmental challenges to the packaging industry are: (1) use minimum or no packaging wherever possible, (2) mandate the use of reusable containers wherever possible, and (3) where nonreusable packaging must be employed, ensure that it is degradable. In order to examine the validity of these packaging concerns and the practicality of the suggested solutions to this apparent problem, one must begin by assessing packaging's true role in litter, solid waste and energy.

Litter

Packaging's role in litter can be ascertained by examination of the data shown in Table 10.2. Packaging products probably comprise about 40% (by number count) of the litter problem. The percentage of packaging in litter is extremely important in understanding packaging's environmental problems. Litter is one of the most obvious forms of pollution. It is visible in all parts of the country to all segments of society, including the well-to-do and other decision makers who might not readily encounter other less obvious forms of environmental upset.

Within the packaging category beverage containers are the largest single indentifiable source of litter. An appreciation of these facts are paramount in understanding the reason for the tremendous environmental outcry against the nonreturnable beverage container. More specific charges against the packaging portion of litter is that it is not biodegradable and "will last forever." This charge was initially hurled primarily at the beverage container portion of litter and, more specifically, at glass and aluminum containers. Degradability seems to be fading as an issue as all parties in the controversy begin to recognize: (1) that nondegradability is an important part of a package's primary function; (2) that degradability happens too slowly to be of much assistance in solving the litter problem; and (3) that the degradation products from a decomposing container could present air or water pollution problems greater than that presented by the littered container.

Recent studies on litter have also emphasized the hazards that litter represents to humans and animals (Syrek 1975). Since many of these hazards are reported to be due to sharp glass or metal items in litter, they can be directly tied to packaging products of one kind or another.

Solid Waste.—The most complete assessment of packaging's role in solid waste management was done by Darnay and Franklin (1969). This report is still the only attempt to assess completely packaging's role in solid waste and it is extremely well done. The report states that packaging wastes made up approximately 19.9% of residential solid waste in 1966. Since packaging waste generation is increasing somewhat more rapidly than the overall solid waste problem, it was further estimated that packaging waste would comprise 21% of the residential waste generated in the United States in 1976. However, while the report gave in great detail the amount of packaging waste, it relied on published reports for the total amount of solid waste being generated each year. As pointed out previously, the solid waste generation figures used in this earlier study appear to be too high based upon more recent analyses. Packaging's role in municipal solid waste is 35 to 40%, rather than the 20% figure reported by Darnay and Franklin. This is important in assessing the direction that environmental pressures will take relating to packaging, and for a greater technical reason perhaps than the litter problem. The importance of finding positive solutions to the solid waste problem cannot be overlooked if packaging is to escape restrictive legislation aimed at curtailing its continued growth.

While (as in litter) the primary relationship between packaging and solid waste appears to be the sheer bulk of packaging that must be disposed of, there have been other charges directed toward specific packages or packaging materials. The question of degradability was raised. However, when it is considered that most solid waste is disposed of in landfills and that most of the problems associated with landfills (leaching of waste materials into the water supply, the generation of explosive methane gas and settling problems) are related to the degradable portion of solid waste, the desirability of degradability in solid waste management is seen to be highly questionable.

One other specific solid waste/packaging aspect warrants mention —the role of plastics. Plastics in general, and halogen-bearing plastics such as PVC in particular, have been the subject of considerable investigation. Except for the degradability issue, virtually all questions regarding plastics and solid wastes have related to incineration. It was said that plastics cause smoke or plastics clog incinerator grates. Regarding PVC the concern was that burning PVC generates hydrogen chloride (a gas) and when and if this gas would combine with water (e.g., in a scrubber) a serious corrosion potential exists due to the formation of hydrocloric acid. In one of the first reports on the subject (Fulmer and Testin 1968), with the exception of the PVC question, it was determined that plastics caused no unique problems in solid waste. Regarding PVC further study was recommended to determine the corrosion and air pollution potential of incineration of halogen-bearing plastics. Subsequent studies (Kaiser and Carotti 1971; Baum and Parker 1972) have indicated that incinerators equipped to handle corrosive gases normally occurring in refuse should be able to incinerate PVC in concentrations that could be expected to exist in solid waste based on current and projected use of PVC and other halogen-bearing plastics. The question has not been completely solved, however. The federal Environment Protection Agency has continually revised questions regarding the incineration of PVC. Recent EPA draft guidelines for handling solid waste at government installations prohibit the use of PVC bags for holding trash although other plastic bags are acceptable. It is not known if this prohibition will be in the final guidlines.

On the other hand, virtually all studies recognize the fact that plastic wastes can be major producers of energy when solid waste is burned. Recent introductions of nonhalogen-bearing plastics into markets with major potential for growth (e.g., beverage containers) coupled with a movement toward energy recovery in most recovery systems (see subsequent section on resource recovery) will be decided plus for plastics in solid waste.

Energy.—Regarding litter it was noted that packaging is an important part of an aesthetically displeasing environmental problem that is obvious and improtant to many of the decision makers of our society. In the solid waste area packaging wastes comprise an important part of a major pollutant—the nation's municipal solid waste problem. However, the technical validity of concern in the energy area suffers by comparison with the former problems. For example, regarding the issue of beverage containers (the most studied aspect of packaging and energy) Hannon's (1973) data shows that a complete switch to refillable beverage containers would save approximately 1/5 of 1% of the nations's energy requirements. Similar results have been reported more recently (Bureau of Domestic Commerce 1975). When one considers that the transportation industry accounts for about 25% of our national energy consumption and space heating for homes and buildings account for 20%, it would appear that efforts in these areas would be more fruitful. Nevertheless, the energy issue is one that is continually discussed in the popular press. The packaging industry is an obvious candidate for attack by those who consider most packaging to be superfluous in the first place.

ALTERNATIVE SOLUTIONS
TO PACKAGING'S ENVIRONMENTAL PROBLEMS

There are a great many approaches to dealing with packaging and the environment. Because of their diversity, each will be discussed with a general framework relating to the specific environmental problem (e.g., litter) being attacked. There will obviously be overlap because some approaches attempt to deal with multiple parts of the problem. Where this overlap occurs it will be identified. It is believed that this format will provide the most logical sequence in putting the various alternatives into contest.

Litter Control

Because it is an extremely visible form of pollution, litter has had the longest history of attempted control.

Promulgation and Enforcement of Antilittering Laws.—Virtually every state has a series of antilittering laws designed to control the littering motorist and pedestrian. Generally the classic antilitter ordinance prohibits the act of littering, which is defined as a misdemeanor under the law, with an appropriate fine of anywhere from $5 to $500. In some instances the antilitter ordinances have been extended to include items such as mandatory litter bags in automobiles or the imposition of a fine on the driver of a motor vehicle from which litter has been tossed. Arrests are rarely made for littering. With only a few exceptions[6] law enforcement officers, judges and society in general have tended to wink at the litterer and the violation of the law goes unpunished. From a practical point of view even strongest enforcement of the litter laws could not directly control the litterer because of the necessity of catching him in the act. However, a firm across-the-board enforcement might do the job. The arrest of one person seen running a red light is often enough to deter most potential offenders, even though one cannot station a policeman at every traffic light. The same type of indirect control would probably also apply to littering.

---

[6] As part of its overall statewide program Oregon has an enviable arrest record for violations of prohibitions against littering.

Keep America Beautiful.—In 1953, well before the current interest in the environment, a group of concerned industrialists met in New York to found Keep America Beautiful (KAB). Top officers of many of America's major packaging companies were included in this group. Their objective was to establish an organization expressly designed for the control of litter. This group has now expanded into a national organization whose sole purpose is to educate the general public against littering and to promulgate a fair but firm enforcement of antilittering laws. The organization has also greatly expanded its base. Its National Advisory Council now includes representatives of a number of environmental organizations as well as representatives of civic and public interest groups. KAB has been criticized in recent years for its industrial history and support. However, this organization recognized the litter problem well before most people and set up positive action programs. Through its system of awards, educational materials and national advertising, the name Keep America Beautiful has become virtually synonymous with litter control.

The Action Research Model.—A recent program called The Action Research Model (ARM) is evaluating the possibility of modifying human behavior in a manner that would make littering recognized as an antisocial activity and hence generate a more favorable climate in the public mind toward litter control.

A psychological research firm, Human Resources Institute, which had developed the concept of the normative systems approach to human behavior, was engaged by a group of KAB supporters to do the research. In this program a systematic approach to litter control is developed within a community. Working first with community leaders and then with all citizens, the littering norms applicable to the community are identified and action programs to change these norms are identified. The basic components of the approach include updated ordinances, modern technology, continuous education and vigorous enforcement of sensible regulations. In the regulatory area the approaches include far more than mere implementation of antilitter laws. Equally important are positive laws, such as those requiring the availability of litter containers, and regulations that the sources of litter (such as garbage receptacles and trucks) be kept covered. The entire interlocking plan is then field tested in cities to ascertain if litter can indeed be reduced by a systematic approach to the problem, including the behavior aspects. It was found that the test cities welcomed the concept not only because it had the potential of controlling the litter problem, which most officials recognize as a minor problem, but even more important, if the approach were successful it could be used to combat more serious forms of antisocial behavior.

TABLE 10.3
MEASUREMENT OF LITTER ACCUMULATION IN ARM DEMONSTRATION
SITES EXPRESSED AS A PERCENTAGE REDUCTION
(BY NUMBER COUNT) IN BASELINE DATA

Principal Sites	
Charlotte, North Carolina	71.6
Macon, Georgia	64.0
Tampa, Florida	58.0
Extension site	
Sioux Falls, South Dakota	51.3

Source: Keep America Beautiful (1975).

The first test programs were started in three cities: Charlotte, North Carolina; Macon, Georgia; and Tampa, Florida. The results after approximately one year of operation (Table 10.3) show truly phenomenal reductions in litter from the base line data taken before the program began. Reductions of over 50% in all litter appear to be achievable by this approach. The successes in the initial test cities have led to expansion of this program to 18 other cities during 1975 with plans for 150 sites by the end of 1976. The ARM is being promulgated under the auspices of KAB and it is anticipated that many other communities will begin this affirmative action program during the next few years.

Legislative Attempts to Control the Litter Problem.—With only a few exceptions, legislative attempts at the local, state and federal levels have been directed at the nonrefillable beer and soft drink container. It is usually required that beverage containers carry a deposit. Other restrictions such as banning the pull tab are possible. Despite the fact that hundreds of attempts have been made to legislate against beverage containers, only Oregon and Vermont have implemented container deposit legislation. In addition, South Dakota has passed a law requiring that only reuseable or biodegradable containers be used after July 1, 1976. California and Minnesota have legislated against the nondetachable end on cans. The state of Washington implemented a model litter control program which is partially supported by a broad-based tax on products considered to contribute to litter.

In eight instances involving state or local situations, beverage container deposit laws have been the subject of referendums. In all cases the proposed deposit legislation has been soundly defeated (Table 10.4). A few local jurisdictions have passed restrictive beverage container legislation, but Oberlin, Ohio is presently the only locality enforcing such legislation.

At the Federal level several pieces of legislation have been

TABLE 10.4
REFERENDUMS—RESTRICTIVE BEVERAGE CONTAINER PROPOSALS

Date	Location	Rejection (%)	Description
November 3, 1970	Washington State	51	Ban on nonreturnable beverage containers. Defeated by a margin of 40,000 votes with one million voting on the issue.
October 5, 1971	Juneau, Alaska	68	Prohibit the sale of beer and soft drinks in containers which do not have a refund value of at least 5¢. 1799 against; 861 for the measure.
May 29, 1973	Bridgeton, Maine	59	Ban the sale of beer and soft drinks in nonrefundable containers. Ordinance passed March 5, 1973. Special election March 29, 1973. 267 for repeal; 180 to retain.
May 21, 1974	Coon Rapids, Minnesota	64	Deposit/ban-the-can ordinance. 2509 against; 1403 for the issue. Ordinance repealed.
November 5, 1974	Crystal, Minnesota	55	Ban the sale of beverages in nonreturnable containers. 3625 against; 2859 for.
November 5, 1974	Dade County, Florida	58	Deposit/ban pull-tab container. 136,016 against; 98,283 for.
April 7, 1975	Ypsilanti, Michigan	65	Ban on nonreturnable, nonalcoholic containers. 2147 against; 1155 for.
October 7, 1975	Eau Claire, Wisconsin	69	Five cent deposit on all beer and soft drink containers. 8244 against; 3714 for.

introduced along the lines of the Oregon law but none has been acted upon. An impartial analysis of the results of restrictive container legislation, such as the Oregon or Vermont experience, is virtually impossible to obtain. It is important to keep in mind that the Oregon statute on beverage containers requires a minimum refund value (a subtle distinction from a deposit). Furthermore, this minimum refund value is $0.02 for refillable bottles that are certified by the Oregon Litter Control Commission as being usable by more than one filler (e.g., the stubby, brown beer bottle). It is $0.05 for all other refillable bottles and all nonrefillable containers. Also, the non-detachable end on the beverage can is eliminated.

The results of the Oregon law 2 yr after passage are confusing. It is generally agreed that litter is down in the state of Oregon. However, the amount of litter reduction, whether or not the litter reduction is due to the Oregon deposit law or other factors, and, most important, the cost of this approach to the state of Oregon is a subject of great dispute. The most definitive study of the situation was conducted by Applied Decisions Systems, Inc. (1974). It was concluded that, overall, litter was down about 10.6% at a cost to consumers, industry and the state of Oregon of $6 to $8 million. The same study showed that beverage related litter was down approximately 66%. Other studies (Gudger and Bailes 1974) dispute the negative economic impact shown in the ADS study. There is only one result of the Oregon law that all sides seemed agreed upon—the nonrefillable beverage container now enjoys only a minute fraction of its former Oregon market. For example, aluminum cans lost approximately 90% of their market since the passage of the Oregon law.

In Vermont the situation is, if possible, even more confusing. The original law which became effective on September 1, 1973 required a simple $0.05 across-the-board deposit on all beer and soft drink containers. Studies under the uniform deposit law (Nadworny 1975) showed a substantial loss in sales for both beer and soft drinks in Vermont. Nadworny reports that in 1974, the first full year of the law, the amount of beer sold in Vermont dropped approximately 11% and the amount of soft drinks sold dropped about 17%. This decline in volume was accompanied by major negative economic impacts in the state of Vermont as well as secondary effects, such as loss of consumer deposits.

In all probability because the deposit law was uniform, cans and refillable glass enjoyed a substantial, although eroding, market as late as 1975. However, in 1975 a new bill was passed in Vermont which requires that glass bottles must be refillable at least five times. It bans nondetachable ends and the plastic ring (Hi-Cone) carrier for cans. This law is scheduled to take effect January 1, 1977.

Regardless of one's position on the beverage container issue, it seems fair to state that the results from Oregon and Vermont must be viewed with extreme caution. Both are relatively low-population states with a highly developed environmental ethic. The results of an acceptance of such legislation in other parts of the United States at this time is certainly speculative. The economic impacts observed in Oregon and Vermont, whatever they may be, would be magnified manyfold in more populat states that represent a proportionally larger share of the national market for beverages and their containers.

Legislation to Regulate Packaging

Legislative approaches attacking beverage containers, while widespread, are recognized by proponent and opponent alike as being extremely narrow. The widespread legislative activity on the beverage container issue, reviewed in some detail in the previous section, in all probability reflects an interest to "do something now," rather than a broad-based approach that would bring the entire packaging industry under some sort of regulation.

The first approach along these lines was the state of Washington's Model Litter Control Act. This act became effective in May of 1971. As part of a broad-based litter control act it imposed a 0.015% business and occupation tax on the gross income of those manufacturers, retailers and wholesalers whose products, including wrappings and containers, contribute to the state litter problem.

A far more broadly based approach was enacted into law in 1973 in the state of Minnesota as part of a bill to "encourage both the reduction of the amount and type of material entering the solid waste stream and the reuse and recycling of materials." This legislation stated that "the (Minnesota Pollution Control) agency shall review new or revised packages or containers except when such changes involve only color, size, shape or printing. The agency shall review innovations including, but not limited to, changes in constituent materials or combinations thereof and changes in closure. When the agency determines that any new or revised package or container would constitute a solid waste disposal or environmental problem, the agency may...prohibit the sale of the package or container in the state." The legislation further stated that the law would not apply to any package or container sold at retail prior to final enactment of the bill (May 25, 1973).[7] After numerous meetings and public hearings, the Minnesota Pollution Control Agency adopted Regulations for Packaging Review on December 20,

---

[7] Minnesota Laws 1973, Ch. 748 Sec. 6.

1974, to be effective on January 1, 1975. In May of 1975 a group of industries began legal action against the packaging regulations. In June of 1975 the temporary restraining order was issued against the Minnesota Pollution Control Agency, restraining it from enforcing the state law and the regulations governing the review of new packages. At the end of 1975 the issue had not yet been resolved.

At the federal level the only legislation actually enacted had been the Solid Waste Disposal Act of 1965 and its successor, the Resource Recovery Act of 1970. This latter legislation was still enforced in 1975 because of several congressional extensions. At the end of 1975 bills that would amend or replace the Resource Recovery Act were in committee. If enacted in the form they were introduced, several would give sweeping powers to the administrator of the federal EPA to regulate packaging, but none would ban or tax any specific product.[8]

For example, S.1744, the Resource Recycling and Conservation Act, which in 1975 was before the Committee on Commerce, would give the federal EPA the power to regulate products based upon their environmental acceptability. This proposed piece of legislation did contain a provision for either house of Congress to disapprove a proposed product regulation.

S.2150, pending before the Committee on Public Works, authorized EPA to prepare and publish a list that includes each type of product, container or packaging technique which inhibits the recovery or recycling of materials from solid waste.

S.1474 would establish "federal packaging guidelines" by requiring the federal EPA to establish guidelines for the packaging of products and publish model standards and regulations to "insure use of types of packaging which best conserve energy and materials. . . ."

It is not known what form the eventual replacement to the Resource Recovery Act of 1970 will take. However, proposed legislation such as that previously mentioned would indicate a

---

[8] In November, 1975, the federal EPA published proposed guidelines that would require all beverage containers sold in federal facilities to carry a $0.05 deposit. In publishing these proposed guidelines, the EPA had used a provision in the Resource Recovery Act of 1970 which required EPA to publish guidelines for "solid waste recovery, collection, separation and disposal systems. . ." as authority for the deposit guidelines. The guidelines relating to beverage containers were but one of a series of guidelines that had been published by EPA during 1975. The others, however, related to a somewhat less controversial area of collection/storage and source separation. At the end of 1975 comments on the proposed beverage container guidelines were being solicited by the EPA and numerous questions had been raised about the authority of the EPA under the Resource Recovery Act of 1970 to issue beverage container guidelines.

reasonable degree of interest to regulate packaging on the part of the U.S. Congress.

Source Reduction

According to the federal EPA: "waste reduction (or 'source reduction') includes a variety of means to control and prevent waste generation through product redesign and change in consumer behavior. It is a unique approach to solid waste management based on the thesis that solid wastes are the unwanted residuals of our production and consumption processes—as are airborne and water-borne wastes—and that the generation of such residuals should be reduced. As presently understood, some of the major choices in production and consumption that are relevant to solid waste generation are: (1) Single use versus multiple use design (2) Shorter versus longer product lifetime (durability) (3) Larger versus smaller products (4) More versus less packaging material per unit of product (5) More versus fewer units of products consumed per family per year (U.S. EPA. 1975)."

What the EPA did not say is that there is a controversy raging between the advocates of source reduction and the advocates of resource recovery, with a great many opting for "business as usual" and not getting too excited about one side or the other. The packaging industry has a vital stake in how this controversy is resolved because its ultimate disposition could have major ramifications to the packaging industry as it is constituted today.

In reality there need not be a conflict between source reduction and resource recovery. Indeed, the objectives of the more reasonable parties on both sides of this dispute are virtually identical. The topic of resource recovery will be taken up in some detail in the next section. However, it is pertinent to lay out some rather profound differences between various approaches to source reduction.

The classic example of source reduction is the elimination of a product or package by legislative fiat. This is the kind of source reduction that is of major concern to the packaging industry. On the other hand, the packaging industry by its very nature has historically practiced source reduction. For example, the metal can is substantially lighter than the glass bottle. The aluminum can (which was introduced later) is substantially lighter than the steel can. Furthermore, all containers have shown dramatic weight reductions as a function of time due to competition between the various container industries. A similar example is in the area of secondary packaging for beverage containers where the Hi-Cone carrier or shrink plastic film for beverage cans is dramatically lighter than the board carrier formerly employed.

Courtesy of Reynolds Metals Co.

FIG. 10.5. EXAMPLE OF NONDETACHABLE END FOR BEVERAGE CONTAINERS

A newer example of source reduction for the packaging industry is the variety of nondetachable ends for the beverage can, which in 1975 were undergoing market tests by numerous companies. Environmental concern coupled with competitive pressures for a better, less controversial package contributed to this development. As outlined by Testin (1972) numerous qualitative guidelines were listed to assist in designing packages for their ultimate disposability.

The basic point is that the ideal package would weigh nothing, take up zero volume and cost nothing, *if* it still performed its basic function. Packagers generally sell to the toughest customers in the world—other businesses who use the package to convey their product to market. Thus competitive pressures alone will tend to drive the packaging industry toward source reduction *within the context of a package's basic function.*

Care must be taken to ensure that pressures from outside the industry for source reduction are considered in implementing new ideas and innovations (such as the nondetachable pull tab end). Yet the packaging industry must be alert to ensure that changes detrimental to product distribution techniques are not imposed from the outside by the unknowledgeable.

Resource Recovery

Of all the solutions to solid waste problems none has caught public imagination more than recycling or resource recovery. It is an approach that virtually all segments of society understand, both from the standpoint of conserving natural resources and of controlling our nation's mounting problems of solid waste. As defined previously, resource recovery in its broadest sense includes the reuse of waste material in any form, whether by direct salvage, conversion of waste materials to new or different products, or the use of the basic energy or chemical constituents of waste.

The concept of recycling or resource recovery (the terms will be used synonomously in this section) is not limited to packaging wastes or to municipal solid wastes. It is an approach that has been practiced for years by the scrap metals industry, and is being used with much success in solving the junk car problem. However, insofar as packaging wastes are concerned, whether their origin is commercial refuse putouts or households, most will end up in the municipal solid waste stream if they are not segregated and recycled before they get to the garbage can. Thus, those areas of recycling of most interest to the packaging industry are: the separation and recycling of used packages before they get to the garbage can, and the extraction of materials in one form or another from mixed municipal solid wastes.

The historic example of sorting and salvaging of waste packaging materials is, of course, in the area of paper. Corrugated and other paper packaging products have always been separated for salvage. The most dramatic new program is the recycling of aluminum beverage cans. Reynolds Metals Company pioneered this approach in the late 1960's in a series of pilot programs designed to develop a method for recycling cans and other clean household aluminum. These ultimately resulted in the present approach in which the public is paid cash for aluminum cans and in some instances for other clean aluminum household scrap. The system is analogous to that used by the waste paper industry.

Reynolds, and later other aluminum companies, established aluminum recycling centers where aluminum could be redeemed for $0.15 per pound ($0.10 per pound until June 1, 1974). There is now a network of over 1,300 collection[9] points where aluminum cans (and often other clean aluminum household scrap) can be redeemed

---

[9] Collection point refers to a location where cans (and sometimes other consumer scrap) are purchased from the public. These are often operated by cooperating beverage companies.

Courtesy of Reynolds Metals Co.

FIG. 10.6. REYNOLDS ALUMINUM RECYCLING PLANT IN WILLIAMSBURG, VA.

for cash. Today Reynolds operates some 25 recycling plants where the cans are received, paid for, processed (magnetically separated and shredded) and shipped to a smelting plant for remelting. In addition the company operates 40 service centers where the public can bring in cans and other clean household aluminum. Some processing (such as compaction) is often employed at these service centers. The company also operates a fleet of mobile recycling units operating on a planned routing of supermarket parking lots, shopping centers, etc., buying cans from the public and servicing collection points. When a current expansion is completed Reynolds will operate approximately

TABLE 10.5
ALUMINUM—CONSUMER RECYCLING PROGRAMS

Year	Aluminum Recycled (millions of lb)
1970	8
1971	34
1972	52
1973	68
1974	103

Source: Testin (1975)

85 permanent recycling facilities and more than 150 mobile units. The success of the aluminum industry consumer program can be seen in Table 10.5. In 1974, 103 million lb of aluminum were recycled.

Nonrefillable glass bottles are also being reclaimed by the glass industry either for glass cullet or for secondary uses such as a substitute paving material.

Steel cans are collected magnetically from numerous municipal refuse and incinerator ash-processing systems. These are used by the steel industry or by the copper industry for precipitation iron. New methods of detinning, new uses for steel cans in the steel industry, the introduction of tin-free steel cans, and the inclusion of magnetic separation in virtually every instance where municipal refuse processing is considered all point toward increased recycling for steel cans.

The most exciting recycling approach for waste packaging materials is the recycling of mixed municipal solid wastes. This is the most general approach. It presupposes no previous separation and considers the mixed solid wastes as an ore to be mined rather than a problem to be gotten rid of.

The national committment to resource recovery is probably most vividly illustrated by the fact that the first federal legislation relating to solid wastes was the Solid Waste Disposal Act of 1965. Its successor was called The Resource Recovery Act of 1970, which was still in force at the end of 1975. Despite the fact that the Resource Recovery Act authorized expenditures of vastly larger amounts than the Solid Waste Disposal Act of 1965 (e.g., over $240 million for fiscal year 1973), expenditures under the Resource Recovery Act were actually quite modest—five to ten times less than that authorized. Nevertheless, the Resource Recovery Act of 1970 was landmark legislation, philosophically, and pointed the nation on the road to resource recovery as a positive alternative to solid waste disposal. Several large-scale resource recovery plants either in operation or currently under construction were funded by monies made available through the Resource Recovery Act.

Courtesy of Black Clawson Fibreclaim, Inc.
Middletown, Ohio

FIG. 10.7. A PIONEER RESOURCE RECOVERY PLANT IN FRANKLIN, OHIO WHERE PAPER FIBERS ARE EXTRACTED FROM REFUSE AND FERROUS METALS, ALUMINUM AND GLASS ARE RECOVERED

In another significant event leaders of major American industry and labor organizations, concerned with the growing national problem of solid waste management, founded in October of 1970 what is today the National Center for Resource Recovery. Originally named the National Center for Solid Waste Disposal, the Center was incorporated in the District of Columbia as a non-profit, non-lobby corporation. Since its inception the National Center for Resource Recovery has become a focal point of resource recovery technology within the United States. In addition, the National Center, in conjunction with a major solid waste processing firm and city of New Orleans, has taken a leadership role in developing a full-scale demonstration resource recovery project for that city.

Courtesy of Union Electric Co.,
St. Louis, Mo.

FIG. 10.8. A 300-TON PER DAY TEST FACILITY IN ST. LOUIS, MO. WHERE REFUSE IS USED TO PRODUCE ELECTRICITY

Finally, and perhaps most important, private industry motivated by profit has taken the lead in developing systems for reclaiming resources from solid wastes. Dozens of companies are now capable of responding to requests for proposal from municipalities and other local governments seeking resource recovery as an alternative to conventional disposal methods.

The concept of resource recovery from solid wastes is simple—its implementation is very complex. Basically the concept consists of taking raw municipal refuse, such as that coming from households and commercial establishments, and processing it through a series of physical, chemical or thermal steps to extract materials and energy. A typical flow chart might include shredding to free the various constituents, their classification to separate the light organics (such as paper) from the heavier material (such as metals and glass). The metals and glass could be further separated by screening to remove the glass particles (which presumable will be shattered in the shredding process), magnetic separation to remove the ferrous metals, and, finally, techniques such as "heavy media" flotation processes to separate the aluminum from the copper and the zinc that remain.

Courtesy of the National Center for Resource Recovery, Inc., Washington, D.C.

FIG. 10.9. A RESOURCE RECOVERY SYSTEM UNDER CONSTRUCTION IN NEW ORLEANS, LA. WHERE PAPER, STEEL, ALUMININUM AND GLASS WILL BE RECOVERED

Such systems are called front end separation systems because they take raw refuse and separate it into its various constituent streams, which may require further processing before ultimately being used by industry. For example, the light organic fraction taken off in the air classifier might be fed into a boiler to generate steam or electricity. Alternatively, raw refuse may be fed into a steam generating incinerator, a pyrolysis unit, or some other device to extract the energy from the refuse. Metals and glass may then be extracted from the resulting ash. A tremendous amount of research and development is required to put these simple concepts into practice. In addition, massive financial and political impediments must be overcome before a resource recovery plant can become a reality for a given area.

As shown in Table 10.6 over 30 resource recovery systems were operating or being planned in the United States at the end of 1975. Resource recovery is not a panacea to solve all of the problems of packaging and solid wastes. Resource recovery systems are currently not economical for small towns or for rural areas with low population density. Furthermore, even if resource recovery plants were in operation everywhere, no plant (except in theory) will recycle everything. There will always be a residual that must be disposed of. However, resource recovery offers a positive alternative to the twin problems of burgeoning solid wastes and dwindling material resources. It is an evolutionary approach that hopefully will be the rule rather than the exception within the next few decades. Despite its limitations, any unbiased analysis of resource recovery will show that it exhibits a far greater potential for reduction in solid wastes than any other approach.

SUMMARY

Packaging's environmental problems relating to litter, solid wastes and energy are extremely varied and complex. No one solution appears to do the job entirely and the threat of outside intervention to control the packaging industry is very real. Progress is being made. Normal competitive pressures will continue to push packagers toward more efficient packaging systems. Items of particular environmental concern (e.g., the pull tab on beverage containers) will probably be eliminated through improved design. Comprehensive programs in litter control (e.g., the state of Washington and the Action Research Model) can make major inroads in the litter problem. Industry-wide recycling programs and resource recovery technology for municipal solid wastes are making real contributions toward solving the solid waste problem.

TABLE 10.6

OPERATING OR PLANNED RESOURCE RECOVERY SYSTEMS
IN THE UNITED STATES[1]

Location	Agency or Company	Type of System	Capacity	Status
Akron, Ohio	City of Akron; Glaus, Pyle, Schomer, Burns and DeHaven[2] (contractor to be selected)	RDF, suspension fired, 100% of fuel steam, for urban heating and cooling and process steam for B.F. Goodrich	1000 TPD	Final designs are completed; bids received February 20, 1975; 90 day holding period on the bid
Ames, Iowa	City of Ames; Gibbs, Hill, Durham & Richardson, Inc.[2]	Materials recovery, refuse-derived fuel to fire three boilers, pneumatically fed in one tangentially oil-fired suspension boiler and two coal fed stoker-fired boilers to produce steam	200 TPD	Startup to begin Spring 1975, expected full-scale operation mid-summer 1975
Baltimore, Md.	City of Baltimore; Monsanto Enviro-Chem Systems, Inc.[3]; EPA; Maryland Environmental Service	Shredding, pyrolysis, water quenching residue, gravity and magnetic separation, and steam generation	1000 TPD (24 hr operation)	Construction completed December 1974; shakedown stage; full-scale operation Fall 1975
Beverly, Mass.	Cities of Beverly, Lynn and Salem, Mass.	Some materials recovery but predominantly energy recovery in the form of RDF or steam for United Shoe Machinery in Beverly who donated the site	500–1000 TPD	RFP will be sent out by July 1975; final selection by August 1975; example of community co-operation with major area industrial firm (United Shoe Machinery)
Bridgeport, Conn.	City of Bridgeport; Connecticut Resources Recovery Authority; Garrett Research and Development Co. (Division of Occidental Petroleum)[3]	Materials recovery: glass, ferrous, and aluminum. Shredded RDF to be marketed to Northeast Utilities Devon Plant for supplemental fuel firing	1800 TPD	Construction expected to be completed mid-1976
Brockton, Mass.	City of Brockton; East Bridgewater Assocs/Combustion Assocs[3]	Ferrous recovery production of Eco-Fuel II® a proprietary form of RDF. This RDF will be fired in an industrial boiler (Weyerhaeuser Corp.)	200 TPD (currently) 600 TPD (mid-1975)	In October 1971, agreement reached between North American Incinerator Corp., CEA, and City of Brockton; plant opened in October 1973, 600 TPD expected by mid-1975

Location	Contractor	Description	Capacity	Status
Chicago, Ill.	City of Chicago; Ralph M. Parsons[2]	RDF to Commonwealth Edison Plant supplement to pulverized coal, suspension fired	1000 TPD	Construction begun, expected shakedown and completion, June, 1976
Cleveland, Ohio	City of Cleveland (contractor to be selected)	Prepared RDF, steam output for usage in city-owned electric utilities	1500 TPD	RFP issued November 1974; bids received on February 5, 1975; decision expected in early summer 1975
Cuyahoga Valley, Ohio	Cuyahoga County Comrs. (contractor to be selected)	Materials recovery (ferrous only) and prepared RDF to generate steam for local industrial markets (steel mills)	1000–1200 TPD	RFP will be issued in May 1975
Erie County, N.Y.	Torrax, Div. of Carborundum;[3] EPA	Torrax pyrolysis system; produces fuel gas	75 TPD	Demonstration program completed; 200 TPD plant now under construction near Brussels to be completed 1976
Franklin, Ohio	Black Clawson Co.[2,3]; EPA; Glass Containers manufacturer Institute	Fibreclaim TM; Hydrapulper magnetic and mechanical separation of water pulped MSW, optical glass sorting, paper fiber recovery, primary sewerage sludge disposal using unrecovered fibers as a fuel	50 TPD (8 hr operation)	Refuse processing since mid-1971; glass color sorting since summer 1973
Harrisburg, Pa.	City of Harrisburg; Gannett Fleming Cordry and Carpenter, Inc.[2]	Incineration in water-wall incinerator of MSW to produce steam	720 TPD	Operating since 1973 at 50% capacity; feasibility study recently completed to supply existing steam distribution system
Haverhill, Mass.	City of Haverhill; Comm. of Massachusetts (contractor to be selected)	Maximum materials and energy recovery, minimum landfill, employing an industrial municipal mix of 25 to 75; no markets provided by RFP	3 sizes: 1000 TPD 1750 TPD 3000 TPD	RFP issued December 1974; bids received March 18, 1975; decision to be announced by Mid-1975
Hempstead, N.Y.	City of Hempstead: Hempstead Resource Recovery Corp.[3] (Div. of Black Clawson/Parsons Whittemore)	Wet pulverizing (Hydrapulper), steam production, possible ferrous, aluminum and glass recovery; organic fraction to raise steam and generate electricity	2000 TPD	Contract signed on December 12, 1974, with Hempstead Resource Recovery Corp.

TABLE 10.6 (Continued)

Location	Agency of Company	Type of System	Capacity	Status
Housatonic Valley, Conn.	Housatonic Valley Connecticut Council of elected officials; Conn. Resources Recovery Authority; Combustion Equip. Associates[2]	Production of Eco-Fuel II®, a proprietary form of RDF used for direct firing into utility and industrial boilers	2000 TPD	Contract awarded June 1974; construction expected to be completed mid-1978
Lowell, Mass.	City of Lowell; Raytheon Service Co.[2,3]; EPA; Comm. of Massachusetts	Incinerator Residue: shredding, magnetic gravity, and other separations from incinerator plant residue	250 TPD (8 hr operation)	Status of construction uncertain due to unforeseen costs to upgrade City of Lowell's incineration
Merrimac Valley, Mass. (Lawrence)	Merrimac Valley Association (contractor to be selected)	Resource recovery, RDF for steam production to municipal wholesale electric corporation for new 100 MW plant adjacent to Plainville site	2000 TPD	RFP issued; bids received April, 1975
Milwaukee, Wisc.	City of Milwaukee; Americology Div., American Can Co.[3]	Materials recovery, ferrous, aluminum, paper and glassy aggregate; RDF prepared as solid fuel for Wisconsin Electric and Power Company which will utilize 600 to 700 TPD of light organic fraction	1200 TPD minimum / 1600 TPD maximum	Contract signed January 1975
Nashville, Tenn.	Nashville Thermal Transfer Corp.[3], I.C. Thomasson and Associates, Inc.[2]	Incineration using MSW as a fuel; designed as a central heating and cooling plant for urban usage	720 TPD initially	Designed to expand to 1500 TPD by 1978
New Britain, Conn.	Connecticut Resources Recovery Authority; SCA Services/ Combustion Equip. Associates[3]	Production of Eco-Fuel II®, a form of RDF used for direct firing into boilers	2000 TPD	Award announced May 1975; completion expected by late 1976
New Orleans, La.	City of New Orleans; Waste Management, Inc.[3] National Center for Resource Recovery, Inc.[2]	Recovery I. Materials recovery (ferrous, fractions, some paper)	650 TPD	Construction begun in November 1974; completion expected 1976

Location	Organization; Contractor	Process	Capacity	Status
Orange County, Calif.	Orange County (contractor to be selected)	Materials recovery and energy conversion	3000 TPD	RFP sent out in February 1975
Palmer Township, Pa.	Palmer Township; Elo and Rhodes, Inc.[2]	Materials recovery, ferrous, non-ferrous, and glass and the refuse fuel to be pelletized for firing with coal in cement kilns	150–200 TPD	Bids for construction have been received
Portland, Ore.	City of Portland; Metropolitan Service District (contractor to be selected)	Ferrous recovery and RDF	2000 TPD	Bids received; final recommendation by May 1, 1975
St. Louis, Mo.	City of St. Louis; Union Electric Co., EPA	Shredding, air classification and magnetic separation; refuse fuel as a supplementary fuel to coal in a utility boiler	325 TPD (8 hr operation)	Operating for several years; program to end early 1976
St. Louis, Mo.	Union Colliery Co. (subsidiary of Union Electric)	Ferrous recovery, RDF preparation for pulverized coal utility boiler supplementary fuel	6000–8000 TPD	Privately owned plant, now under construction, expected to be completed in June 1977
San Diego County, Calif.	San Diego County; Garrett R & D Co.[2,3], EPA	Shredding, air classification, drying, mechanical separation, pyrolysis, magnetic separation, froth flotation. Pyrolytic oil to be used as a supplemental fuel	200 TPD (24 hr operation)	Construction to be completed mid-1976
Saugus, Mass.	City of Saugas; RESCO (joint venture of De-Matteo Constr. and Wheelabrator-Frye)[2,3]	Incineration in water-wall incinerator of MSW to produce steam; steam to be accepted by nearly industrial user (General Electric, Lynn plant)	1200 TPD	First boiler on line in September 1975 with second boiler on line in June 1976 for a total capacity of 2400 TPD
South Charleston, W. Va.	Linde Division, Union Carbide Corp.[3]	Purox[TM] oxygen converter (pyrolysis); high temperature partial oxidation using pure oxygen, no feed preparation needed	200 TPD	Demonstration plant owned by Union Carbide Corp.
State of Delaware	State of Delaware; EPA	Shredding, air classification, magnetic separation, sewage sludge disposal, and refuse fuel as a supplemental fuel for oil in a utility boiler	500 TPD	Contractor selection expected mid-1975

TABLE 10.6 (Continued)

Location	Agency or Company	Type of System	Capacity	Status
Toledo, Ohio	City of Toledo: Ralph Parsons[2]	Ferrous recovery, shredding, air classification, and RDF preparation; expect to employ RDF as supplemental fuel with pulverized coal in utility boiler	1200 TPD	RFP expected to be issued June 1976
Washington, D.C.	District of Columbia; National Center for Resource Recovery	Shredding, air classification, magnetic separation, aluminum eddy-current separation, glass froth flotation, and RDF preparation for supplemental fuel undensified and densified	240 TPD	Test Facility in various stages of operation; aluminum recovery mid-1975, glass recovery, early-1976, RDF utilization early-1976

[1]This listing is not necessarily complete.
[2]Denotes engineer.
[3]Denotes contractor.

Source: National Center for Resource Recovery (1975).

The key question as to whether or not outside intervention to control the packaging industry for environmental purposes actually occurs is probably one of timing. If legislators and other decision makers are convinced that environmental problems attributed to packaging are being solved rapidly enough, then the probability of control will be greatly lessened. On the other hand, if progress is not deemed to be occurring at a sufficiently rapid pace, restrictions, in one form or another, for all or part of the packaging industry would be most likely.

Therefore, it is incumbent upon the packaging industry to press forward as rapidly as possible with improved packaging to lessen packaging's solid waste and litter impact, to reduce packaging's material and energy consumption (within the confines of the package's basic function) and to participate actively in alternative positive solutions to packaging's environmental problems (such as resource recovery).

BIBLIOGRAPHY

APPLIED DECISION SYSTEMS. 1974. Study of the Effectiveness and Impact of the Oregon Minimum Deposit Law. Legislative Fiscal Office, Salem, Oregon.

BATTELLE COLUMBUS LABORATORIES. 1972. A Study to Identify Increased Opportunities for Increased Solid Waste Utilization. National Association of Secondary Materials, Inc., New York.

BAUM, B., and PARKER, C. H. 1972. Plastics Waste Disposal Practices in Landfill, Incineration, Pyrolysis, and Recycle. Manufacturing Chemists Association, Washington, D.C.

BUREAU OF DOMESTIC COMMERCE. 1975. The Impacts of National Beverage Container Legislation. U.S. Dept. Commerce, Washington, D.C.

BINGHAM, T. H., and MULLIGAN, P. F. 1972. The Beverage Container Problem, Analysis and Recommendations. U.S. EPA Rept. *EPA-R2-72-059.*

DARNAY, A., and FRANKLIN, W. E. 1969. The Role of Packaging in Solid Waste Management 1966 to 1976. Publ. *SW-5c.* U.S. Dept. Health, Education and Welfare, Washington, D.C.

FINKNER, A. L. 1969. National Study of the Composition of Roadside Litter. Report from the Highway Research Board of the Division of Engineering, Natl. Res. Council, Natl. Acad. Sci.—Natl. Acad. Eng. to Keep America Beautiful, Inc., under subcontract *HRB-88-69-3.*

FULMER, M. E., and TESTIN, R. F. 1968. Report on the Role of Plastics in Solid Waste. The Society of the Plastics Industry, Inc., New York.

GEORGE, P. C. 1970. The CMI Report on Solid Waste Control. Communications Marketing, Inc., Washington, D.C.

GUDGER, C. M., and BAILES, J. C. 1974. The Economic Impact of Oregon's "Bottle Bill." Oregon State Univ. Press, Corvallis, Ore.

HAGERTY, D. J., PAVONI, J. L., and HEER, J. E., JR. 1973. Solid Waste Management. Van Nostrand Reinhold, New York.

HANNON, B. 1973. System Energy and Recycling: A Study of the Beverage Industry. Center for Advanced Computation, Urbana, Ill.

KAISER, E. R., and CAROTTI, A. A. 1971. Municipal Incineration of Refuse with 2% and 4% Additions of Four Plastics: Polyethylene, Polystyrene, Polyurethane, and Polyvinyl Chloride. The Society of the Plastics Industry, Inc., New York.

KEEP AMERICA BEAUTIFUL. 1975. Action Research Model Bull. *5*, New York.

NADWORNEY, M. J. 1975. Some Economic Consequences of the Vermont Beverage Container Deposit Law. Univ. of Vermont, Burlington, Vt.

NATIONAL CENTER FOR RESOURCE RECOVERY. 1974. Resource Recovery from Municipal Solid Waste, A State-of-the-Art Study. Lexington Books, Lexington, Mass.

NATIONAL CENTER FOR RESOURCE RECOVERY. 1975. Technical Manual for the Pennsylvania Solid Waste-Resource Recovery Development Act. Dept. of Environmental Resources, Penn.

SORG, T. J., and HICKMAN, H. L., JR. 1970. Sanitary Landfill Facts. U.S. Dept. Health, Education and Welfare, Public Health Service, Washington, D.C.

SYREK, D. B. 1975. California Litter. A Summary. Principal findings abstracted from a comprehensive study report prepared for the California State Assembly, Committee on Resources and Land Use. Institute for Applied Research, Carmichael, Calif.

TESTIN, R. F. 1972. Disposal of waste packaging materials. *In* Principles of Package Development, R. C. Griffin, Jr., and S. Sacharow (Editors). Avi Publishing Co., Westport, Conn.

TESTIN, R. F. 1975. Recycling Opportunities and Challenges for the Aluminum Industry. Pres. Am. Chem. Soc. Symp. Energy and Materials, Washington, D.C.

TESTIN, R. F., and DROBNY, N. L. 1970. Processing and Recovery of Municipal Solid Waste. J. Sanit. Eng. Div., Proc. Am. Soc. Civil Eng. 7345, SA 3.

THE SOCIETY OF THE PLASTICS INDUSTRY. 1970. The Plastics Industry and Solid Waste Management. The Society of the Plastics Industry, New York.

THE UNIVERSITY OF CALIFORNIA PACKAGING PROGRAM. 1969. Proceedings of the First National Conference on Packaging Wastes. Co-sponsored by Bureau of Solid Wastes Management, U.S. Public Health Service, Packaging Industry Advisory Committee. Univ. of California, Davis.

TRAIN, R. E. 1974. Win the War on Waste. Pres. Third Natl. Congr. Waste Management Technol. and Resource Recovery, San Francisco, Calif.

U.S. CODE ANNOTATED. 1969 Suppl. Solid Waste Disposal Act of 1965. Title 42, Public Health and Welfare, U.S. Govt. Printing Office, Washington, D.C.

U.S. EPA. 1974. Office of Solid Waste Management Programs, Second Report to Congress, Resource Recovery and Source Reduction, Publ. *SW-122.*

U.S. EPA. 1975. Office of Solid Waste Management Programs, Third Report to Congress, Resource Recovery and Waste Reduction, Publ. *SW-161.*

U.S. 91ST CONGRESS. 1970. 2nd Session, House of Representatives. Rept. *91-1579.* Resource Recovery Act of 1970. U.S. Govt. Printing Office, Washington, D.C.

Index

Acids, 128
Acrylics, 80
Acrylonitrile butadiene styrene (ABS)
    resin, 80
Acrylonitriles, 58, 160
Alginates, 164
Aliphatic hydrocarbons, 157
Alkali cellulose, 109
Alkyds, 80
Alumina, 52
Aluminum, 3, 31, 44–46, 50–51, 56, 58–61,
    99, 172, 175, 232
    cans, 22, 226, 228
    corrosion resistance, 59
    emissivity, 58
    gas resistance, 60
    grease and oil resistance, 60
    heat and flame resistance, 60
    moisture vapor, 59
    opacity, 60
    thermal conductivity, 59
    workability, 59
Antimony, 42
Arsenic, 42–43

Bags, 1, 101
Bakelite, 66, 68, 70
Bauxite, 51–52
Benzyl alcohol, 141
Beryllium, 43
Bismuth, 42–43
Blackplate, 49
Blister packages, 113
Blow molding, 77
Bronze, 134
Burlap, 1
Butadiene styrene, 80
Butcher wrap, 101
Butter wrap, 101

Cadmium, 43
Carbon dioxide, 116
Carothers, 62–63
Casein, 80
Cellophane, 4, 12, 66, 108–111, 148, 184
Celluloid, 111, 184
Cellulose acetate, 67, 80, 111–113
Cellulose acetate butyrate, 80

Cellulose nitrate (celluloid), 65–66, 68, 70,
    80
Chemical pulp board, 93
Chipboard, 96
Chromium, 44
Chromium coated steel, 46
Chromium oxide, 50
Chromium trioxide, 67
Citric acid, 133
Clay, 106
Closures, 30–39
    collapsible tubes, 31–33
    containers, 30–31
    glass, 27–28
    screw, 33–37
Cloth, 1
Coated litho papers, 106
Cold rolling, 54–55
Collagen films, 163
Color, 11
    in glass, 18–19
    in plastics, 73
Compression molding, 76
Container glass, 23–26
Copolymers, 142, 144–145
Copper, 41–42, 46, 232
Corrugated board, 96–98
Cotton, 183
Cotton linters, 111
Crystal, 120, 126
Cylinders, 105

Dehydrated soups, 102
Diamines, 148
Die lines, 112
Dimethyl phthalate, 112, 143

Ediflex, 162
Energy, 214, 218, 232
Epoxies, 80
Ethyl cellulose, 80, 197
Ethyl glycol, 140
Ethylene dichloride, 147
Ethylene vinyl acetate, 103, 128
Extruded film, 112

Fabricated sheet packages, 56
Flexible packaging, 99

Fluorocarbons, 80, 158
Foods, dehydrated, 9
  meat, 8
Fourdrinier, 105
Freezer paper, 101

Gallium, 41, 46
Galvanized steel, 49
Glass, 3, 232–233
  borosilicate, 20
  composition, 18–21
  container manufacture, 24–27
  history, 17–18
  lead, 19
  photosensitive, 20
  properties, 20–22
  silica, 20
Glassine, 103, 106
Glycol, 129
Gold, 41, 45–46, 99
Gum rubber, 122

Homopolymers, 124, 142
Hosiery, 103
Hydrocarbons, 123, 133
Hydroxpropyl cellulose, 163

Incineration, 212
Indium, 46
Ingots, 52–54, 171–172
Injection molding, 77
Intaglio, 201
Ionomers, 115, 121
Ions, 203
Iridium, 45–46
Iron, 41, 47

Kaolin, 106
Klucel, 163
Kraft paper, 100–103
  board, 94

Lamination, 78–79
Landfill, 211, 213, 217
Lead, 41, 61–62
Litter, 209, 211, 213–214, 216–217, 219–
  220

Machine finishing, 196
Machine glazing, 105
Magnesium, 44–46

Malaysian tin, 49
Manganese, 43–44, 46–47
Mercury, 41
Metals, 41–63, 233
Methyl chloride, 132
Microcrystalline, 196–197
Mineral oil, 143
Molten pig iron, 47
Molybdenum, 44

Nickel, 41, 43–44
Nip rolls, 155
Nitriles, 81
Nitrocellulose, 108, 184
Nylon, 5, 81, 199–200

Oil, 174
Ozone, 303

Packaging, 215
  areas of, 8, 10
  definition, 5
  education, 12–14
  industry, 206
  products, 216
  role in waste, 317
Palladium, 45
Paper, 2, 84–89, 99, 102–103, 148, 232
  coating, 89
  production, 85
Paperboard, 3
  cartons, 114
  containers, 32–33, 99
  types, 89–90
Paraffins, 196
Pectins, 164
Pewter, 46
Phenol, 120
Phenoxies, 158
Phosphate-chromium, 50
Plasticizers, 71–72, 112, 120, 135, 143, 151,
  153–154, 156
Plastics, 3, 6–7, 65–83, 218
  chemistry, 68–70, 74
  fillers, 71
  for closures, 31
  formulation, 70–74
  history, 65–68
  processing, 75–79
  specific gravity, 70
Platinum, 43, 45–46

Pliofilm, 99
Pollution, 206, 212, 219
    air, 218
Polyallomers, 82, 122
Polyamides. *See* Nylons
Polycarbonates, 82, 145–147
Polyesters, 82, 137–138, 140–141, 159, 165, 199
Polyethylene, 5, 82, 99, 101, 113–115, 117, 122–123, 138, 147–149, 159, 165, 197, 199
Polyolefins, 189, 202–203
Polyphenylene oxide, 158, 165
Polypropylene, 82, 122–124, 126–128, 186, 199
Polystyrene, 82, 128–131, 159, 167, 185
    foam, 132
Polyurethane, 82, 158
Polyvinyl alcohol, 164
Polyvinyl chloride, 67, 82, 150, 153, 156, 159, 167, 188
Polyvinylidene chloride, 82, 108, 142, 167
Porosity, 100
Propionates, 82
Protective coatings, epoxy, 50
    oleoresinous, 50
    phenolic, 50
    vinyl, 50
Pulp, 84
    production, 85–89
Pyrolysis unit, 233

Recycling, 229
Resin, 103, 112
Rhodium, 45
Rubber, 100, 183

Sacks, 1
    paper, 100
Saran. *See* Polyvinylidene chloride
Shrink films, 126
Silicon, 46–47, 82

Silver, 41, 46, 99
Sodium bicarbonate, 133
Steel, 4, 46, 48–49, 55
    plate, 49
    rolls, 105
Sulfate paper, 101
Sulfur trioxide, 202
Sulfuric acid, 111, 202
Supercalendered paper, 105

Temper, 55–57
Terneplate, 49
Tetrahydrofuran, 145
Thallium, 46
Thermoforming, 77
Tin, 4, 41, 48–49, 62
Tin-free steel, 50, 55
Tinplate steel, 46, 48–50, 55
Tinplated sheet iron, 46
Tissue paper, 103–104
Titanium, 44–45
Toluene, 135
Transfer molding, 76
Tube metals, 60–63
Tubular films, 120
Tungsten, 44

Unbleached kraft, 100–101
Urea-formaldehyde, 109
Urea resins, 82

Vanadium, 44
Vinyl acetate, 121

Wadding, 104
Wax paper, 102–103
Wood, 3, 101
    hard, 101
    shaft, 182
    soft, 101

Zinc, 42, 44, 46, 232